The landscape of the SUSSEX DOWNS

Area of Outstanding Natural Beauty

A man looking down from the crest of the Downs to the south and to the north of him sees much of what his ancestry have seen since men first stood upon those hills. The Weald was once a little denser in wood, the coastal plain a little less thick with villages, but that is all. The high, broad belt of the sea has always made a frame for that view. The flooded river valleys have always picked it out with patches of silver. The roll of the Downs has always stood, like a monstrous green wave, blown forward before the south-west wind. The simple and vivid green of the turf, and the sharp white chalk pits, have always stood making the same contrast with the sky and the large sailing clouds.

Hilaire Belloc, The County of Sussex, 1936

John Tyler

The landscape of the SUSSEX DOWNS

Area of Outstanding Natural Beauty

Text Prepared by Landscape Design Associates
for the Countryside Commission and the
Sussex Downs Conservation Board.

First published in 1996 by

Countryside Commission
John Dower House
Crescent Place, Cheltenham
Gloucestershire GL50 3 RA

Sussex Downs Conservation Board
Chanctonbury House
Church Street
Storrington
West Sussex RH20 4LT

ISBN 0 86170 466 5
CCP 495

This book was designed by the Sussex Downs Conservation Board.
Set in Adobe Garamond, Gill Sans and Bodoni Poster.
Printed by The Beacon Press, on sequel satin recycled, a totally chlorine free paper, using a new environmentally friendly process that eliminates the use of water and alcohol.

Cover Photograph: Martin Page *View towards Firle Beacon From Ditchling Beacon*
Photograph this page: John Tyler

Contents

Foreword

The Sussex Downs is an outstanding landscape of lowland England. For centuries, the chalk hills, river valleys and sandy heaths of Sussex have provided work and shelter for local people and refreshment for visitors. But the pressures of modern life are such that, if our successors are to enjoy the landscape we love, we must identify what it is that makes it special and how we can work together to ensure that its special qualities are protected and, if possible, enhanced.

The purpose of this report is to capture the special qualities of the Sussex Downs, identify how the landscape has come to be as it is, what the special characteristics are of the various landscape types which together make up the whole picture, how the landscape relates to the ecology of the area and the pattern of human settlement, and how the landscape has been perceived by writers and artists. The paper examines the forces for change influencing the Sussex Downs and concludes with an assessment of the importance of the landscape of the AONB.

It is, as the report concludes, a quintessentially English landscape. It may not be wild or remote like many of our National Parks, but the homely and pastoral landscape of the Sussex Downs has its own special appeal, summing up for many of us the best of lowland England. This report, which follows others on AONBs around the country, is the first to get to grips with the nature of the Sussex Downs landscape in ways which will be helpful to all those - politicians, planners, administrators, countryside managers, farmers and landowners, conservation and access groups - who love the Sussex Downs and want to see the essential character of the area maintained and enhanced.

We commend this report to these groups, to the general reader and student, and to all those who share our love of the Sussex Downs. The publication of this document is an essential step towards securing the conservation of this most important feature of our national heritage. The report is complementary to the management strategy for the AONB which is being published simultaneously by the Sussex Downs Conservation Board and, taken together, these two documents will be of vital importance for the future management of the Sussex Downs AONB.

Lord Nathan
Chairman
Sussex Downs Conservation Board

Richard Simmonds
Chairman
Countryside Commission

John Tyler

Introduction

Your Britain · fight for it now. The South Downs by Frank Newbold

The Sussex Downs are the south-east spur of the South Downs, extending from Hampshire to the coast at Eastbourne. The chalk is tilted, with a prominent north-facing escarpment and a dip-slope to the south. Directly to the north, and sandwiched between the chalk uplands of the North and South Downs, lies the Weald, part broad lowland vale and part rugged greensand ridges. To the south of the Sussex Downs, the heavily populated coastal plain narrows progressively from a broad plain near Chichester to a tiny strip at Brighton, before disappearing completely as the chalk uplands meet the sea at the coast between Brighton and Eastbourne.

The chalk downland of Sussex and a portion of the western Weald were designated as an Area of Outstanding Natural Beauty (AONB) in 1966. (The western part of the South Downs was designated the East Hampshire AONB in 1962.) This means that the scenic beauty of this landscape is recognised to be of national importance, a resource worthy of protected status.

The Sussex Downs are renowned for their scenic beauty. The rolling chalk uplands, precipitous escarpments and dramatic chalk cliffs along the coast at Beachy Head are magnificent examples of chalk scenery. To the north, the wooded greensand ridges, hidden ghylls and ancient pastures of the scarp footslopes and the Weald have a more mysterious, secretive quality. This is a contrasting area of a deeply rural unspoilt landscape abutting a large urban area to the south. The area has a special distinctive identity and strong contrasts in landscape character. It is steeped in history, with ancient hill-forts, barrows, Roman roads and deserted medieval villages representing the long continuity of human influence. Centuries of traditional land management practices have resulted in a rich diversity of natural habitats, including flower-studded chalk grassland, ancient woodland, flood meadow, lowland heath and the particularly rare chalk heathland.

The AONB contains a truly stunning range of landscapes which have inspired generations of artists, writers and visitors.

The landscape suffers from the consequences of its popularity and is subject to constant pressures for change. It is essential that conservation measures are carefully targeted to counteract changes which would result in degradation of the landscape while also conserving and enhancing the distinctive qualities which contribute to its scenic beauty.

To this end, the Sussex Downs Conservation Board has adopted a strategic approach to the conservation and management of the AONB, concentrating on policies which are 'landscape-led'. It is important that the landscape is researched, recorded and understood. This study is intended to provide a starting point and framework for this approach. (It is supplemented by a wide range of more detailed specialist reports, many of which are referenced in the bibliography.) The study begins, in Chapter 2, with a summary of the physical and human history of the landscape, from the formation of the underlying bedrock to the Ice Age and from the arrival of the first nomadic hunters and gatherers to the recent impact of the combine harvester and the golf course. This is followed by an analysis of the landscape types which are found within the AONB. Based on a more detailed landscape assessment undertaken by the Countryside Commission and the Sussex Downs Conservation Board, Chapter 3 provides a catalogue of the AONB landscape, concentrating on the special qualities which make it distinctive and thus providing the baseline knowledge necessary for its conservation and enhancement.

Chapter 4 describes the diverse range of habitats found within the AONB and the traditional systems for landscape management which have established and maintained them. This is followed by a section which focuses on the ways in which the AONB landscape has been perceived and interpreted by the many artists and writers whom it has inspired: Chapter 5 traces this process, from the sixteenth century historical accounts to the present day.

Chapter 6 describes the dominant forces for change within the AONB. The special qualities of the landscape are at risk from changes in agriculture and forestry, from development pressures, infrastructure and mineral extraction and from recreation and tourism. While such change cannot be halted, it should be understood and carefully guided to ensure that any damaging effects are minimised. There may often be scope for landscape enhancement.

The final chapter summarises the key components of this valuable landscape heritage and the measures being taken to ensure that it is conserved for future generations to enjoy.

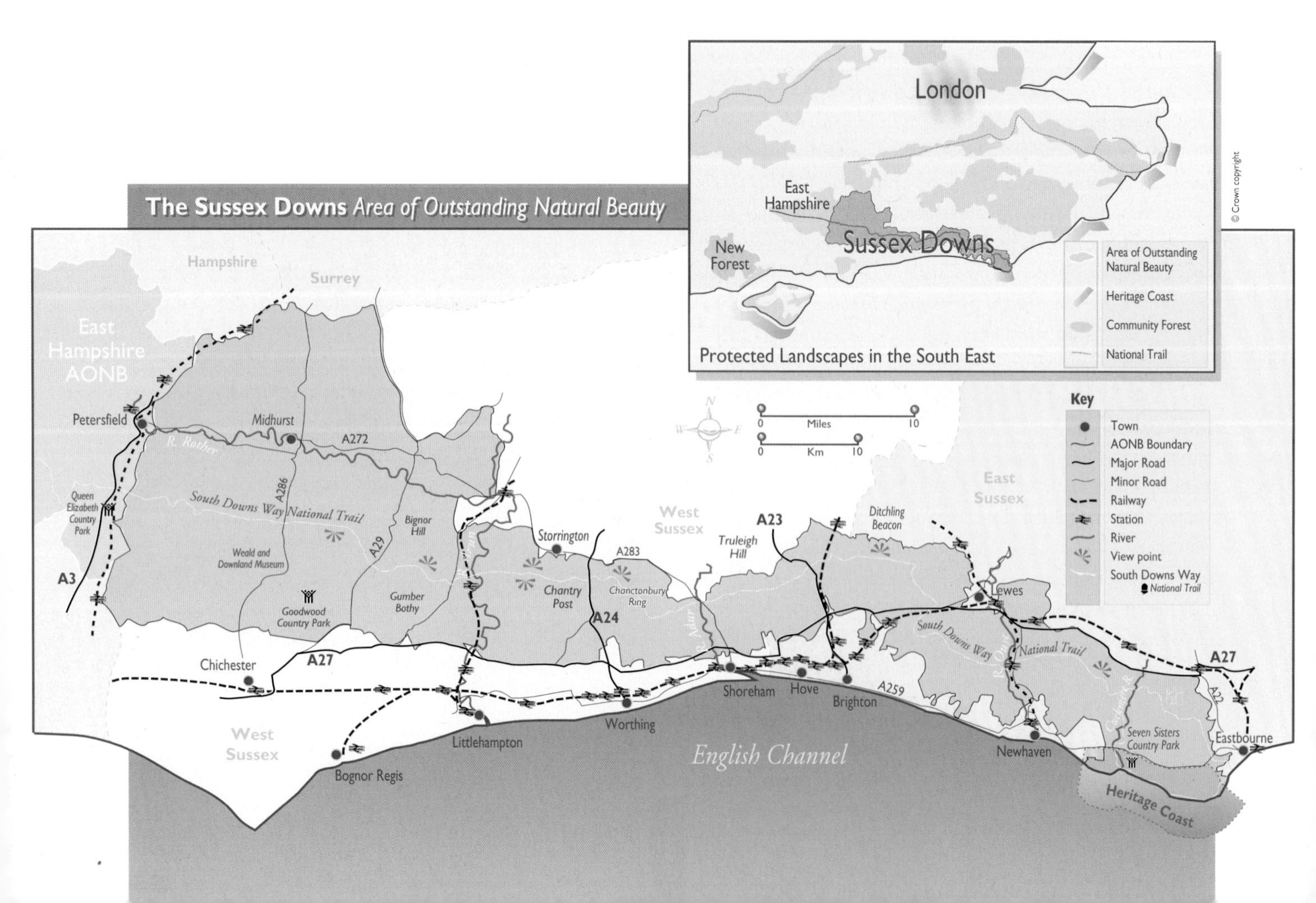

The Shaping of the Sussex Downs Landscape

John Tyler

The Seven Sisters from Cuckmere Haven

Physical Influences

Approaching the Sussex Downs from the north, the chalk escarpment looms as a dramatic, fluted wall abutting the gentle plain of the Low Weald. Yet this impressive range of hills is an eroded remnant, a mere relic of a cataclysmic event which took place 20 million years ago.

Subterranean pressures, built up over vast periods of geological time, resulted in massive movements of the Earth's crust. The area which is now south-east England was then a low plain, formed from layers of chalk, clay, silt and sand, continuously deposited and compressed over the Cretaceous period. The upheaval which created the Alps pushed these layers of rock upwards to form a vast rounded dome, centred on the area which is now known as the Weald.

Gradually, over the course of millions of years, the land changed shape. The centre of the dome was eroded away, leaving an outer upstanding rim of chalk surrounding a lowland plain formed from older layers of clay and sandstone. The outer rim of chalk forms the uplands of the North and South Downs and the central plain is the Weald.

The underlying structure of the North and South Downs reveals the original form of the wealden dome. The layers of chalk are tilted at a sharp angle, with steep scarp slopes facing inwards, towards the centre of the original dome, and shallow dip-slopes leaning towards its outer edge. The older clays and sandstones of the Weald have been eroded to form a complex range of landforms.

These areas of chalk and sandstone form some of the finest and most evocative landscapes in southern England. The Sussex Downs AONB includes the long, curving spine of the South Downs, together with the sandstone ridges and clay vales of the

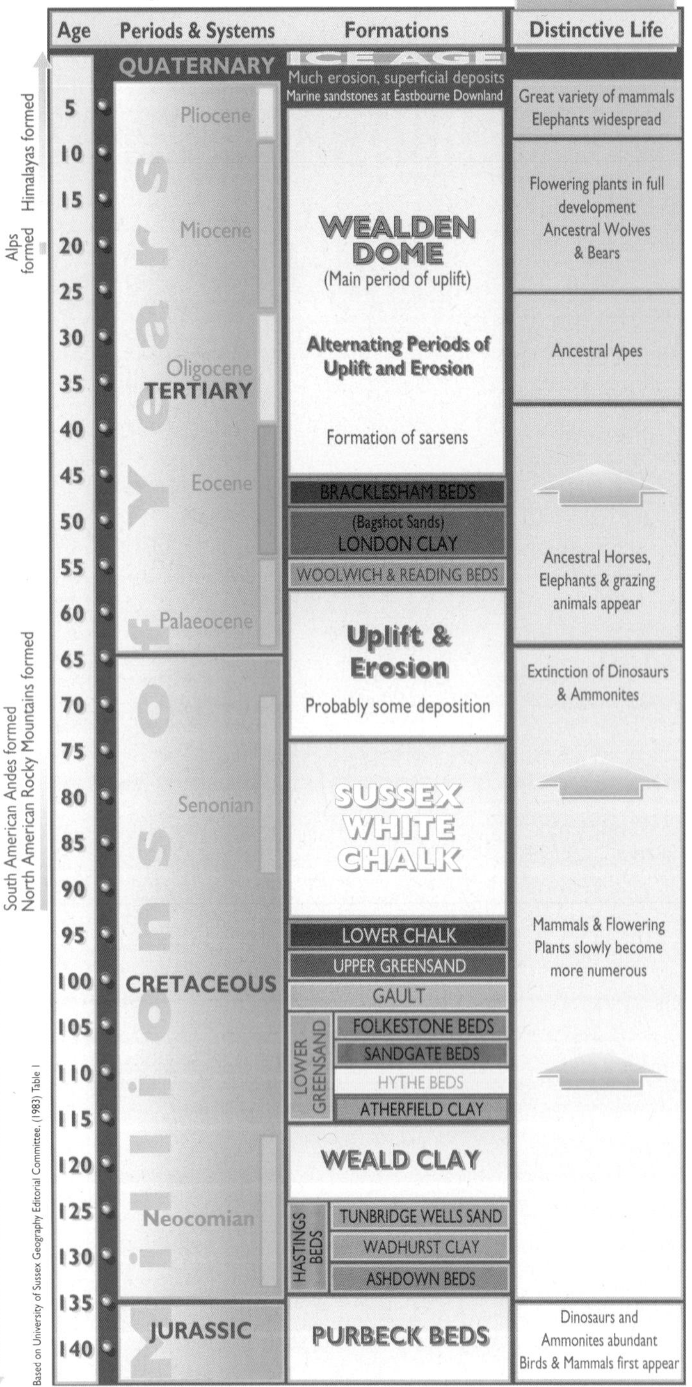

western Weald. It is a varied landscape of stunning contrasts and immense diversity, a legacy of its complex geological structure and the influence of centuries of human activities.

Geology, soils and landform

The key to understanding the AONB landscape is provided by its underlying rocks and soils. Their configurations determine not only the physical characteristics of landform, local climate and vegetation cover, but also the way in which the landscape has been used and adapted as a resource by wildlife and people alike.

The uplift and progressive erosion of the wealden dome has exposed a fascinating sequence of geological history, from the older rocks towards the north of the AONB and the original centre of the dome, to the younger formations of the chalk downland to the south. The diverse scenery of the AONB landscape results from this varied, underlying structure and, in particular, differences in the resistance of the various rocks to the processes of weathering and erosion.

The oldest rocks in the AONB are the clays and sandstones of the western Weald, exposed by the progressive erosion of the wealden dome. These layers of clay, silt and sand were deposited 130 million years ago during the early Cretaceous period when south-east England was a low-lying landscape of shallow freshwater lakes, deltas and marsh. Alternate layers of sand and clay were formed as water levels in the lakes fluctuated, producing undulating clay lowlands drained by a branching network of minor streams.

The Mid-Cretaceous period 110 million years ago was marked by the inundation of the marsh and mud-flats on which the wealden clays had accumulated. A warm sea covered most of southern England and the majority of the rocks structuring the AONB landscape were formed from layers of marine sedimentation on the sea-bed during the next 40 million years.

The Hythe Beds of the Lower Greensand were some of the first sediments laid down on the sea-bed and are particularly resistant to weathering. They form the elevated, steeply undulating relief of the north-west Weald. The poor soils are heavily wooded with patches of heathland and rough grazing in open glades. Narrow streams have cut deep ghylls in the sandstone, which rises to a steep, horseshoe-shaped escarpment sheltering the clay lowlands on the margins of the Weald. The escarpment reaches a maximum height of 280 m (920 ft) at Black Down, the highest point in the AONB.

To the south of the Hythe Beds, there is a sequence of Lower Greensand rocks. The Sandgate Beds form rolling relief with well-drained, easily eroded sandy soils which are almost exclusively used for arable farmland. Further south again, the Folkestone Beds form a slightly elevated, flat-topped plateau which is associated with poor soils and extensive tracts of heathland. The structure of the Folkestone Beds shows evidence of strong underwater currents which produced enormous ripples of sand across the sea-bed. The deep seams of sand are valuable, and as a result, the heathlands of the Folkestone Beds are pitted with sand-pits and quarries.

During the Lower Greensand period, the sea gradually deepened and the waters became still. Under these conditions

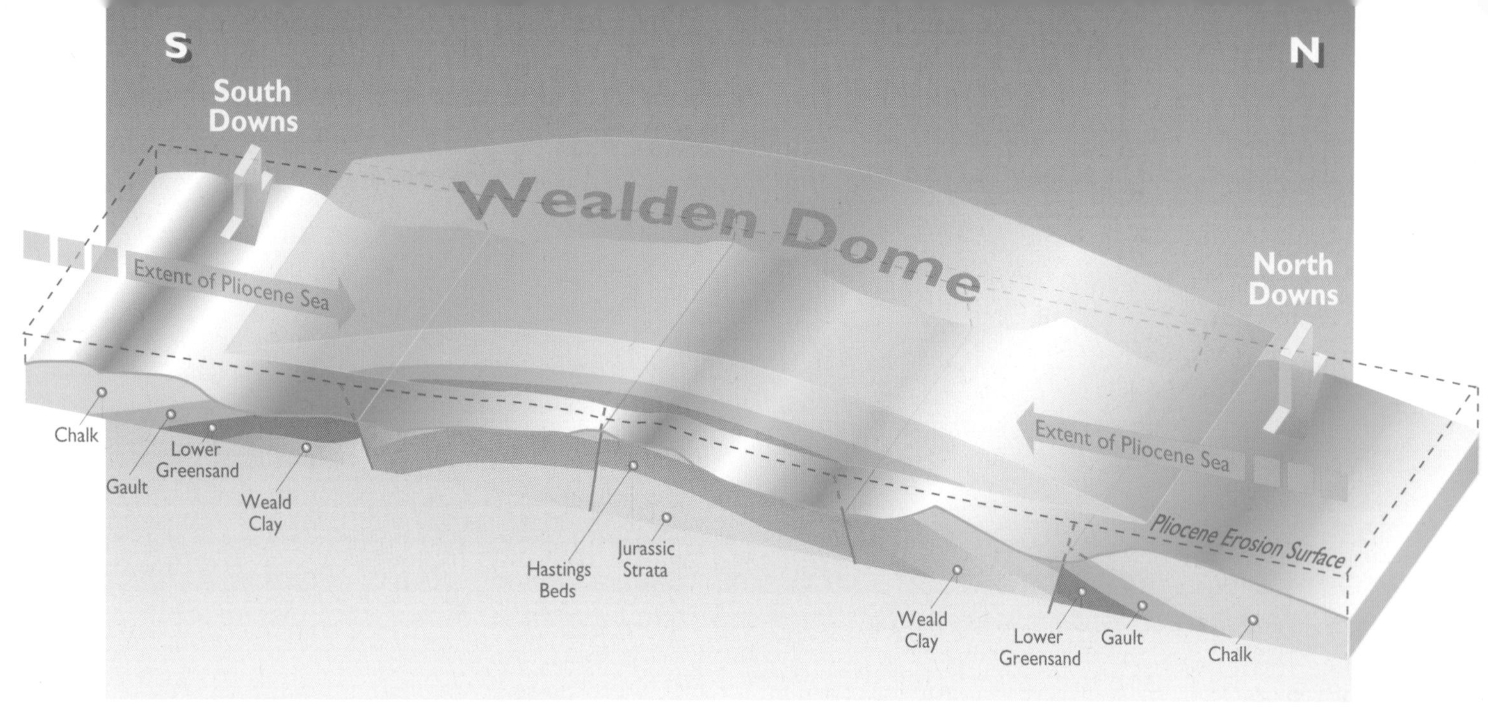

Cross section of the wealden dome and chalk escarpments

a thick, dark mud accumulated. This now forms the Gault Clay which lies next in the sequence of Lower Greensand beds and which forms a clay vale to the south of the heathland.

At a later date, strong underwater currents were again established and, in West Sussex, the sandier sediments of the Upper Greensand layers were deposited. The rock produced from this layer is relatively resistant and forms a low bench at the foot of the chalk escarpment towards the western boundary of the AONB. Further east, the underwater currents were evidently less active since there is no bench of Upper Greensand and the Gault Clay lowlands extend right up to the foot of the escarpment.

The AONB landscape is dominated by the chalk escarpment, formed from the youngest rocks in the geological sequence. Chalk originates from the skeletal remains of marine creatures deposited, compressed and eventually fossilised on the sea-bed to form layers of pure, hard limestone. It is a relatively homogeneous rock and most parts of the South Downs are structured by the formation known as Upper Chalk, which is characteristically embedded with flints. In some areas, the chalk bedrock is overlain by more recent, superficial deposits. The most significant, Clay-with-Flints, is a reddish-brown sandy clay deposited during the Tertiary period and usually found in patches on the more elevated parts of the chalk dip-slope.

The simple structure of the chalk uplands is strikingly revealed in profile where the escarpment is abruptly truncated by the sea at the glistening white cliffs of Beachy Head. Just to the west, the famous hanging valleys of the Seven Sisters provide a dramatic cross-sectional view through one of the dry chalk valley systems so typical of the dip-slope.

The chalk forms expansive, rolling upland relief. There is little surface drainage, but the combination of structural folding, differential erosion and the effects of the Ice Age has produced uneven slopes, carved into sweeping forms with coombes on the scarp and extensive dry valley systems on the dip-slope. Towards the margins of the dip-slope, a line of hills and ridges forms an

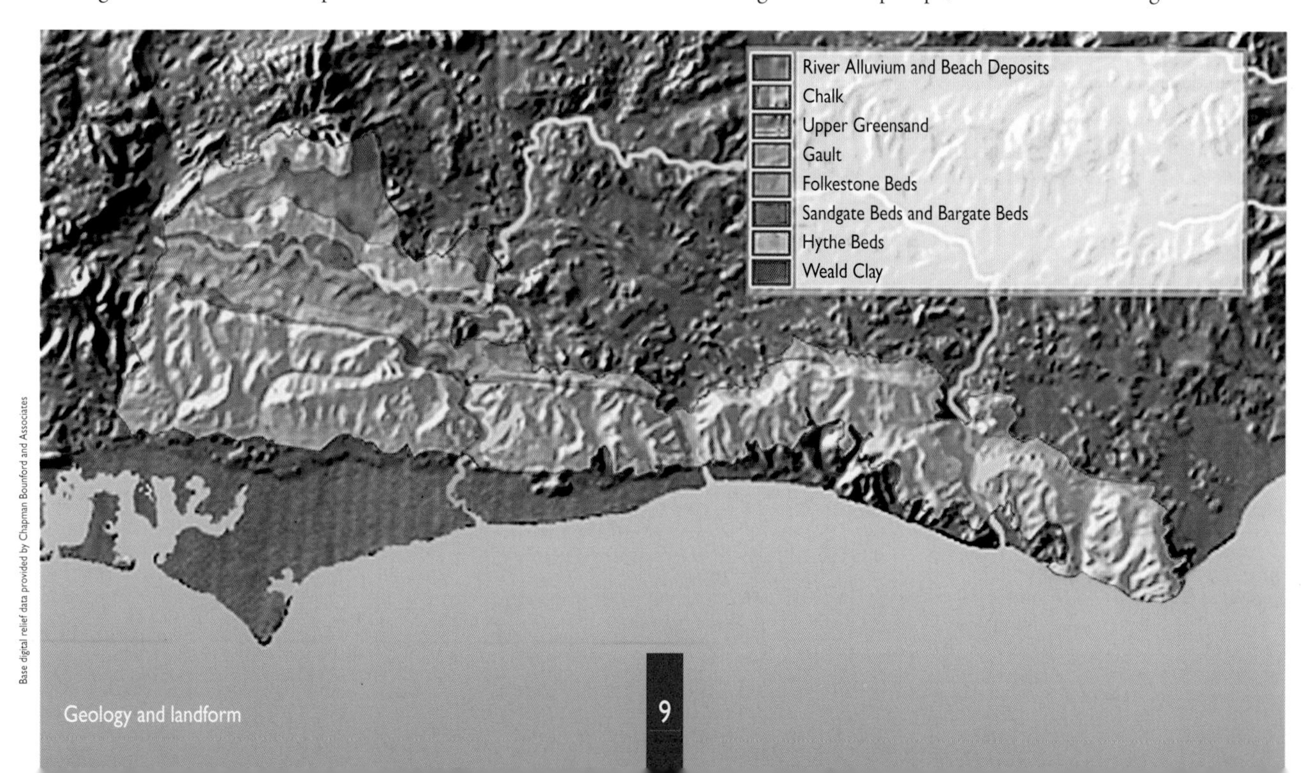

Base digital relief data provided by Chapman Bounford and Associates

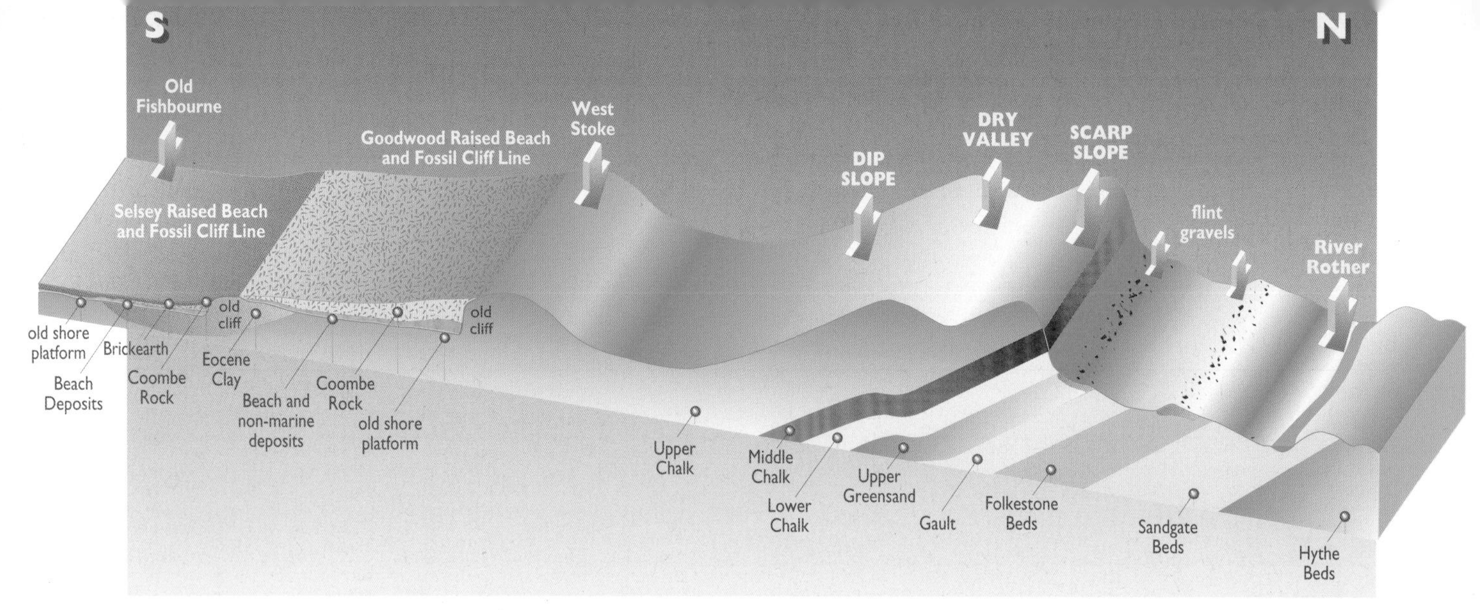

Physical structure of chalk: escarpments, dip-slope and valleys

intermittent secondary escarpment, thought to result from variations in the resistance of the different types of chalk which outcrop on the dip-slope. The South Downs are a textbook for geomorphologists, with numerous striking examples of typical chalk scenery. Some of the most dramatic landforms, such as the Devil's Dyke, an exceptionally steep coombe which slices into the scarp at Poynings, have been the subject of much debate amongst the experts. It seems that such strong erosive effects may have resulted from the enormous shifts in temperature and hydrology associated with the Ice Age.

The Sussex Downs are subdivided by several principal valleys, the largest of which contain the rivers of the Arun, Adur, Ouse and Cuckmere. These four rivers have eroded deep, U-shaped valleys right through the chalk, creating direct north-south links between the Weald and the coast. The extensive floodplains of these rivers provide a strong contrast to the dry valleys and expansive arable farmlands of the surrounding chalk downland.

Changing sea-levels

The southern boundary of the AONB hugs the margins of the Sussex Downs, only including a small but spectacular portion of coastline where the chalk ends in the sea. However, as recently as 8,000 years ago the sea-level was much higher and the coastline reached the lower chalk dip-slope, several kilometres inland from its present position.

Ancient fossil cliff-lines and shore platforms, carved into the dip-slope, provide evidence of this former coastline. Sand and shingle were deposited at the foot of the cliff and this raised beach can be traced today in the Goodwood area of West Sussex.

During the Ice Age which followed, the sea-level was lowered as ice formed over much of the continent. This lower sea-level caused the principal rivers to erode with renewed vigour, particularly in their lower reaches, towards the coast. The steep valley sides of the lower Arun valley, near Arundel, and the Ouse valley to the south of Kingston were eroded at this time.

As the ice melted and the sea-level rose again, these lower river valleys became tidal estuaries and marine deposits can be found underlying the more recent alluvium on the floodplains as far north as Bramber and Lewes.

Human Influences

For centuries the landscape has been used as a resource and patterns of human activity have shaped its character, each phase building on the pre-established pattern. At times, evidence of earlier activities and type of land use have been erased by subsequent developments, but throughout the AONB there are numerous examples of places where early patterns of farming, settlement and industry are preserved. These archæological and historical features enhance local identity and sense of place, often on a minute scale, and provide a further dimension to our experience of the landscape.

From prehistory to the Romans

The earliest settlements were on the chalk downland, which is littered with important archæological sites. Most summits are encircled by the ramparts of Iron Age hill-forts and ancient burial mounds and the remains of stock compounds and flint mines all provide evidence of the activities of early man. It comes as no surprise to find that the earliest human remains in the country, dating from palaeolithic times, have recently been unearthed at Boxgrove on the chalk dip-slope of West Sussex.

The Boxgrove site dates from around 500,000 BC and evidence suggests that nomadic gatherers roamed the area for thousands of years, fishing, foraging for food in the woodlands, and hunting deer and wild boar. By 6000 BC, Mesolithic man was chipping the local flints embedded in the chalk to make tools and arrow heads. These settlers probably migrated from one temporary camp to the next, but are likely to have planted some seeds and made simple arrangements for food storage.

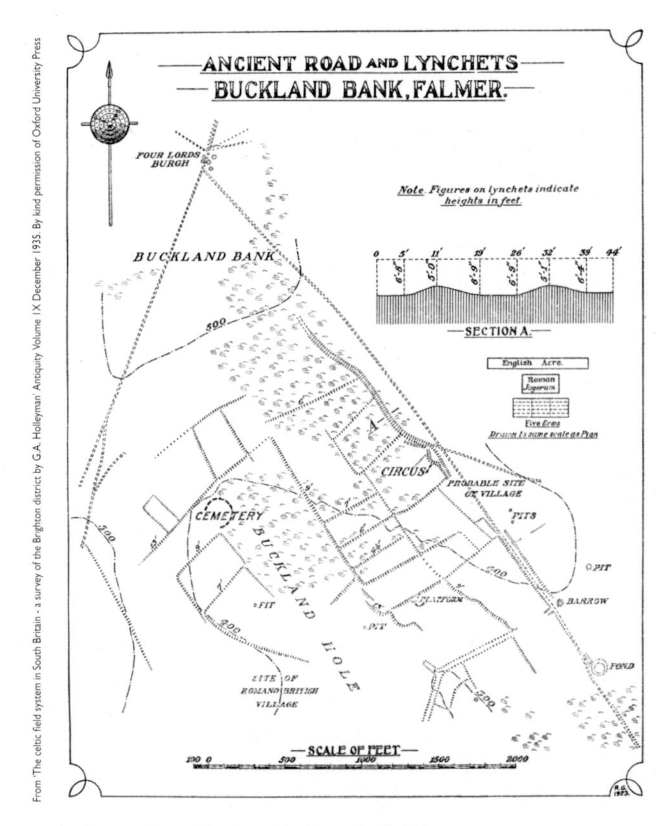

Ancient roads and lynchets Buckland Bank, Falmer

Neolithic man (4500-2000 BC) is generally credited with the development of early forms of agriculture. There is evidence for the development of stronger flint tools, more suited to clearing trees and tilling the soil. The light chalk soils would have provided ideal sites for the first relatively permanent settlements and a more organised, specialised culture was established. The earliest ancient barrows and flint mines discovered so far are associated with this period. Cissbury Ring was an important centre for flint mining in Neolithic times, rivalled only by Grimes Graves in Norfolk.

From about 2000 BC onwards (Bronze Age), a new tribe known as the Beaker Folk came to Sussex. They were responsible for introducing the first metal tools, made of bronze, and for subsequently clearing much of the woodland on the chalk downland. There is evidence that massive soil erosion occurred on the chalk slopes at this time, a direct result of such tree clearance. The downland was used for grazing stock as well as for cereal cultivation.

The beginning of the Iron Age is marked by a deterioration in the climate as conditions became colder and wetter. Settlements became larger and pressures on agricultural land increased. The introduction of iron tools, particularly the plough, together with population pressures, led to the rapid expansion of agricultural activity on the heavier soils of the Weald. Woodland was cleared and farmsteads were established throughout the AONB.

Early Celtic field systems on the Downs were typically small, geometric fields bounded by lynchets. Many have been reduced to mere bumps by constant ploughing, although the site at Buckland Bank, Falmer has been partially spared.

The most impressive relics of this period from a landscape point of view, are undoubtedly the great hill-forts. Perched on the most prominent summits, these defensive sites must have been designed to protect defined territories. They may have been treated as meeting places and occasional safe refuges for the surrounding, relatively scattered population. Many, like the Trundle in West Sussex, were built on the site of earlier Bronze Age settlements. The steep ramparts and ditches were supplemented by wooded palisades and the hill-forts must have become increasingly important as the arrival of the Romans threatened established territorial rights.

In his paper *Man in the Downland Landscape* (1994), Peter Brandon writes:

> *By the beginning of the Roman occupation from 43 AD it is thought that the woodland clearance was almost total and that arable cultivation had then reached a 'high-water mark' on the Downs that was not exceeded until after the Second World War.*

This intensification of arable farming was associated with a relatively stable government and a rapid population expansion. There was a proliferation of towns and villages, together with new roads and luxurious country villas, such as at Bignor.

Artists impression of the Roman Villa at Bignor

Sussex Downs
Historical Table

Millennium

Age	Periods	Events	Dates
	20th	**Sussex Downs Designated AONB**	1966
		National Parks & Access to Countryside Act	1949
		2nd World War	1939-45
		1st World War	1914-1918
	POST MEDIEVAL	Repeal of Corn Laws	1846
		General Enclosure Act	1801
		French Revolution	1789
		First Sussex Turnpike Trust	1749
1536		Civil Wars *(England)*	1642-1649
		Spanish Armada	1588
		Columbus sails for the New World	1492
	MIDDLE AGES	**Wealden Iron Industry** *(Blast Furnace)*	1490-1540
		Black Death	1348-1349
		Battle of Lewes	1264
		Magna Carta	1215
		Domesday Survey	1086
1066	ANGLO SAXON (in England) Dark Ages	Norman invasion of Britain	1066
410		**Sussex conversion to Christianity**	681
	ROMAN (in England)	Construction of Hadrian's Wall	123
43 AD		Roman invasion of Britain	43
	IRON AGE (in England)	Birth of Christ	AD
BC		Construction of Hillforts	BC
		Great Wall of China	214
750	BRONZE AGE	First cities developed *(Middle East)*	
		First accurate calendar devised (Egypt)	
		Beaker Folk in Sussex	2,000
2,400	NEOLITHIC PERIOD	Ancient barrows	
		Cissbury Ring flint mines	
		Great Pyramid in Egypt	2,700
		Stonehenge	2,800
4,500	MESOLITHIC PERIOD	Early flint tools used	
		Walls of Jericho	
		First pottery in Far East	
10,000	PALAEOLITHIC PERIOD	Earliest cultivated plants S.E Asia	
		Migrant tribes *(through Europe & Britain)*	
	500.000yrs	Boxgrove Man	

The development of a Roman military town at Chichester would have been a strong catalyst for local production and trade. The Romans encouraged the expansion of the Celtic iron industry, building a network of roads to link quarries and furnace sites with urban centres.

It is likely that this part of south-east England was particularly peaceful and prosperous under the Romans as a local Romano-British chieftain, Cogidubnus, surrendered at an early stage and offered the Romans support and the opportunity to establish a local base at Chichester. Stane Street, a Roman road built directly across the Downs and the Weald linking Chichester and London, would have been a metalled, cambered road up to 7.5m wide - an impressive example of the Romans' engineering achievement.

From the Anglo-Saxons to the medieval period

The withdrawal of the Romans in the fifth century left an economy in a state of partial collapse and a society which proved extremely vulnerable to attacks by Saxon marauders. The earliest Saxon sites were on the chalk downland between the Ouse and the Cuckmere rivers, but their influence spread rapidly.

A general decline in population led to the collapse of the wealden iron industry and a reduction in cultivation in some of the more marginal areas. The western chalk uplands were used only for sheep pasture, with crop production concentrated in the valleys. There was generally an emphasis on stock rearing rather than cultivation in Saxon times, an indication of a less prosperous economy. Extensive droves were developed for the passage of sheep from the downland to the wealden pastures and tracks were hollowed out by the constant trampling of livestock. The urban centres became less significant and rural settlements, particularly those along the foot of the Downs, were favoured as farmsteads in this area: they had access to both the lighter soils of the chalk and the heavier clays of the Weald and could thus maximise their efficiency. Mixed farming was the norm and dairy cattle were fattened on the brooks pastures of the Ouse, Arun and Adur.

The present place names of the majority of wealden villages suggest they were established by this period. Those in the Low Weald were likely to have been small hamlets surrounded by areas of wood-pasture for cattle, pigs and sheep. Farming in this area was supplemented by employment in the iron industry and its associated woodland management.

In AD 681, Sussex was converted to Christianity by Bishop Wilfrid. This event also marked the establishment of a system of land division by charters. Thereafter, land grants could be documented and the system of dividing land into major units known as Rapes and smaller Hundreds (groups of early parishes) was established. During the late Saxon period, towns, markets and ports again became more important, largely because they were relatively easy to defend in the face of Viking raids. Rapes were centred on important towns, the burghs, which were fortified against the raiders.

The Norman conquerors ousted the Saxon barons from their estates and established a rigid feudal system. They built powerful stone castles, such as that at Arundel, to guard the principal river valley communication routes. The Domesday

Book, produced for taxation purposes, provides an invaluable record of people's possessions and lifestyles at this time. It indicates that rural settlements in the early medieval period were typically small and widely scattered. Roads tended to follow a casual, meandering alignment, linking individual settlements by the most convenient route. This was a feudal society and an 'open-field' system of farming was well established. The large communal fields surrounding the hamlets were divided into strips, each farmed by an individual. This was again a period of relative prosperity and population growth and there is evidence of agricultural expansion, with forest clearance, cutting and grazing on heathlands and marshland drainage.

This expansion came to an abrupt end in the mid fourteenth century with the arrival of the Black Death. Many villages, such as Exceat and Hangleton, were deserted and the ordered open-field farming system fell into disarray as farming communities had a surplus of land at their disposal and there was a retreat from more marginal land.

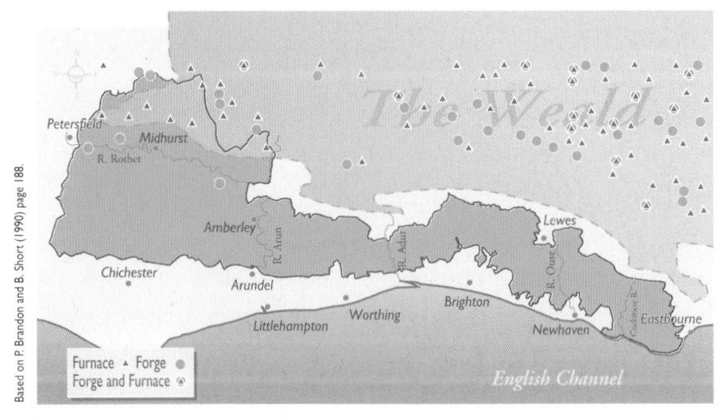

Based on P Brandon and B. Short (1990) page 188.

Plan of wealden forges and furnaces

The wealden iron industry

The ironmakers of the early medieval period continued to practise their relatively primitive manufacturing techniques. The early industry was concentrated in the High Weald, where iron ore deposits were found near the junction between the Wadhurst Clay and the Ashdown Sand. Wood, required to produce the charcoal, was plentiful and the practice of coppicing was widely established to ensure a steady supply of the younger stems most suited to charcoal production.

In the west, iron ore was found near the top of the Weald Clay, where the iron ore was dug from deep mine pits. The ground worked for these mine pits often retains a pock-marked, uneven surface, with slight, rounded craters indicating the location of the original shafts.

The industry expanded rapidly with the introduction of the early blast furnace, an invention introduced by French migrants. Water power was used to work the bellows for the blast furnaces and hammers for the forges, and the fast-flowing streams running off the sandstone ridges were dammed to create hammer ponds with sluices to control and harness the flow. The manufacture of glass also became important in some areas.

Iron making began to decline from the late sixteenth century onwards. The market for pig iron, the most important product, had become insecure and water supplies were often unpredictable. Coke was increasingly substituted for charcoal and other parts of the country were able to provide more cost-effective, large-scale production. The ponds were taken over by local corn mills and the coppices were used for fuel, charcoal and construction.

The production of iron in the Weald was never on a massive, efficient scale and the industry remained essentially rural, but the area was nevertheless the first recorded industrial centre in the country.

Landed gentry

The rise of the Tudors brought a wave of prosperity to the countryside as feudalism wilted and estates were bought up by a new class of wealthy landowners. The favoured spot was the strip of land to the north of the chalk escarpment, a particularly fertile agricultural area where villages clustered along the spring-line. Some of the new gentry were wealthy ironmasters; others had made their money from trade or through crafty allegiance to the Crown. Most lived in the manors and great houses they owned and contributed to the prosperity and management of the surrounding countryside.

Parham House (built 1577), Wiston House and Glynde Place are magnificent examples of such Tudor mansions and Cowdray (completed in 1540) was grander still. A hundred years later, the Civil War left many such houses in ruins and Royalist lands in disarray, but this period of Tudor landed wealth has left a permanent imprint on the landscape.

Glynde Place

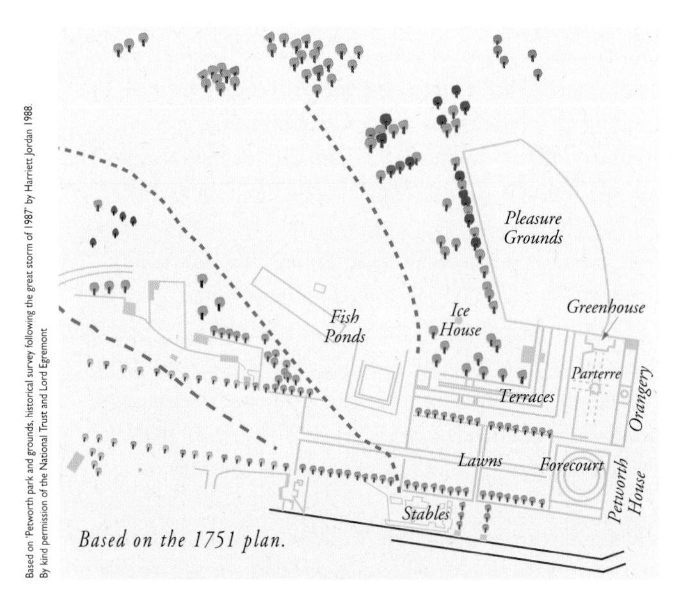

Petworth Park - showing the formal layout in 1751

Agricultural prosperity in the eighteenth century brought a further phase of wealth to the land-owning elite. The great houses and estates at Petworth, Goodwood, Uppark and Stansted all reflect this prosperity. The fashion was to clear the picturesque and earlier formal terraces and gardens to make way for new landscaped parks. The work of Capability Brown at Petworth Park sparked off a flurry of activity. Exotic trees such as cedars of Lebanon were introduced and planted in carefully designed groups to frame views, lakes were dug and 'ruined' follies constructed to lead the eye into the landscape and to provide a destination for designed, circuitous walks. At Parham, an entire village was moved so as not to spoil the view from the house.

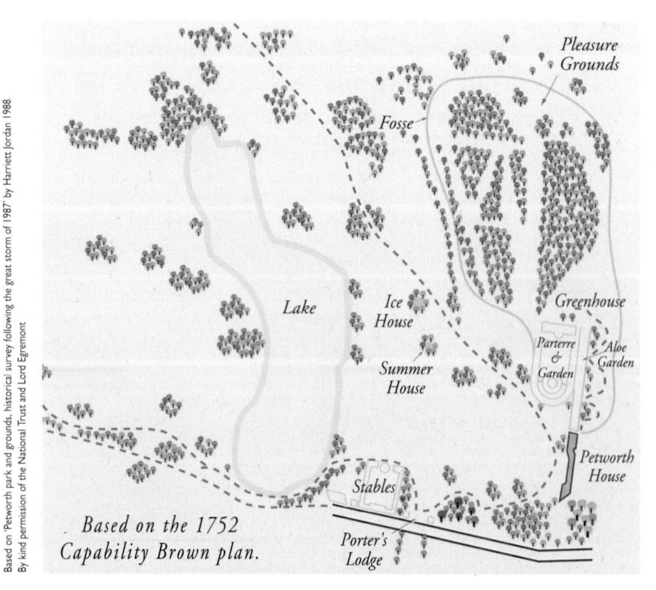

Petworth Park - Capability Brown's landscaped design, 1752

Agricultural change

During the eighteenth and nineteenth centuries, agriculture was transformed, although the new crops, tools, techniques and rotations associated with the 'agricultural revolution' would have been introduced sporadically and over a long period of time. Much depended on the attitude and prosperity of local landowners. Many of the rich landowners were practical farmers who encouraged agrarian improvements. Chalk pits were dug to provide lime for the wealden farms; waterways, such as the Rother and the Arun, were 'improved' by a series of cuts, canals and locks for better navigability. Lord Egremont established a model farm at Petworth and supported many improvement projects in West Sussex.

Enclosure was the first stage in the process. It was often resisted by the smaller farmers, for whom it meant a loss of grazing rights. From the seventeenth century onwards the countryside was gradually reorganised from an open-field system into the patchwork of fields, hedgerows and woodlands familiar today. The new fields were often irregular, reflecting local agricultural conditions and, sometimes, the earlier strip-field pattern. In places the frontier of cultivation extended into the less hospitable landscape of the heaths and forests, leaving the remaining unenclosed 'waste' and commons as the last refuge for the commoners to kill rabbits, graze their animals and gather fuel. For instance, the heathlands to the south of the western Rother were divided between several parishes along the southern side of the valley.

There was a rapid increase in sheep farming, with sheep grazing on the open chalk downland by day and folded at night. The river floodplains were subdivided into a patchwork of ditches controlled by dams and hatches, designed to drain the wetter valley meadows.

The downland was farmed for both sheep grazing and arable crops with the proportion of land under crops varying according to market conditions. The traditional pattern involved grazing in the river valleys, arable on the lower slopes and rough grazing on the upper slopes. High prices for cereals led directly to an increase in arable land and there is evidence that ploughing led to the decline of high quality chalk pastures.

W H Hudson, writing at the end of the nineteenth century, describes the disparity between the original chalk greensward and areas reclaimed by the plough:

But not all the untilled downland is turf. There are large patches of ground, often of twenty or thirty to a hundred acres in extent where there is no proper turf, and the vegetation is of a different character. Some of these patches are of a very barren appearance, and others are covered with flowers in Spring but in Summer are dry and yellow-brown ... This is due to the fact that the ground in some former period has been ploughed for periods of five to five and twenty years ... This land spoilt by the plough is said by the shepherds to be 'cickly' and the grass that grows on it, little in quantity and poor in quality, they call 'gratton grass'.

It was largely a question of productivity and the more sheltered, deeper soils of the chalk dip-slope were generally favoured for arable land.

From about 1875 until the Second World War, agricultural prices were depressed and much arable land was replaced by scrubby pasture. The ploughing campaigns of the war years reversed this trend (Brandon, 1994):

By the spring of 1942, the East Sussex Agricultural Committee had reclaimed over 8000 acres of former turf-covered downland for wheat and yields had been outstandingly successful. It was estimated that this land and further land on the Downs scheduled for reclamation would produce annually enough wheat bread for 240,000 persons in addition to cattle food, capable of producing 2,400,000 gallons of milk annually, together with fresh vegetables for the coastal towns and a sugar ration for 125,000 persons.

In addition, the War Department requisitioned some 22,000 acres (8,900 ha) of the downland between Littlehampton and Eastbourne for military training. This spared the downland from further ploughing. Some of the metalled roads up on to the Downs date from this period of military occupation, notably the road between Steyning and Sompting and the roads to the summits of Beddingham Hill and Firle Beacon. The traditional links between the Weald and downland agricultural regions were also severed. When farming was re-established after the war, the old mixed pattern of traditional downland farming, which utilised a range of landscape types from the chalk uplands to the valleys and brooks pastures, was lost. Despite these ravages, the traditional image of grazed downland became increasingly significant. The Downs became a symbol of national pride and patriotism during the war years, when posters depicting the distinctive rolling hills under a sunny sky reminded the country of its homeland.

Changing patterns of recreation

The London-to-Brighton railway was completed in 1841 and led to a booming leisure industry as Londoners flocked to the seaside resorts along the South Coast, and to Brighton in particular. This was the heyday of the Royal Pavilion and the racecourses at Brighton and Lewes. The Downs themselves were a popular, convenient day trip and Devil's Dyke became a special attraction, with a railway to take visitors on a round trip from Brighton.

Throughout the twentieth century, the rural landscapes of the South Downs and the Weald have increasingly been valued for their distinctive scenic beauty. The proximity of London has added to the popularity of the area, but at the same time to the pressures for change. Early planning legislation and public land ownership offered some protection, particularly in the face of the rapid expansion of towns along the coastal plain and on the clifftops of the East Sussex Downs. The designation of the Sussex Downs AONB in 1966 was intended to protect the scenic beauty of this nationally important landscape, while also enhancing it as a setting for quiet recreation. Today the landscape of the AONB is valued not only by its residents, but also by millions of visitors, attracted by its history, wildlife and beautiful scenery.

Nevertheless, the Sussex Downs AONB is not a fossilised landscape. The great houses have been forced to 'diversify' to support their estates. Glyndebourne has its opera season and Goodwood its racecourse and there is a scattering of golf courses throughout. Statutory planning restrictions ensure that urban development is no longer such a threat, but road construction continues to slice up the countryside.

Garland Photograph Collection - print no. N37865 By kind permission of West Sussex County Council Archivist

The old Roman Road over the South Downs above Bignor - *postcard*

John Tyler

Goodwood Racecourse

Changing agricultural and forestry policies are an important influence as landowners are increasingly encouraged to manage their land more sympathetically.

This brief history of the landscape indicates above all that it is a dynamic resource. The landscape we know today has been shaped by the interplay between the physical influences of geology, landform, climate and human activities. The geological timescale is incomprehensibly vast and human influence represents a mere scratch on the surface, following millions of years in which the landscape was formed by nature alone.

But the scale of human influence is out of all proportion to its relatively short history and the pace of change continues to escalate. For most people, creeping urbanisation, recreational demands and the intensification of agriculture are seen as detrimental influences. The Sussex Downs are a case in point but, by its designation as an AONB, there is an opportunity to slow and influence the pace of change in this nationally important landscape.

Change is not necessarily a bad thing, but there is a need to understand its cause so that it can be guided constructively. By 'reading' the landscape of today it is possible to understand its complex history and the ways in which the Sussex Downs AONB might continue to change in the future. The following chapter, in which the AONB landscape is described and analysed in relation to its component landscape types, provides a basis for this approach.

From Branch lines to Midhurst' by Vic Mitchell and Keith Smith. By kind permission of Middleton Press. Other albums by Middleton Press covering railways that give or have given access to the Downs are:- 'Branch lines around Midhurst'; 'South Coast railways -Brighton to Eastbourne'; 'Southern main lines - Crawley to Littlehampton', 'Haywards Heath to Seaford', 'Three Bridges to Brighton'; 'Kent & East Sussex waterways', 'London to Portsmouth', waterways, 'West Sussex waterways'.

Angel Hotel advertisement

Variations in Landscape Character

John Tyler

Looking over the Weald from Harting Down

Introduction

Geology is the key influence on landscape character, although soils, natural processes, land use and the cultural history of human activities all combine to create a diverse range of landscape types within the Sussex Downs AONB.

A landscape assessment of the Sussex Downs AONB, undertaken in 1994, has provided a new descriptive map of the area and a visual analysis of the special features and components of the landscape and their relationship to one another. The aim was to identify and describe the aspects of the AONB landscape which make it so distinctive and special.

The landscape assessment divided the AONB landscape into fifteen landscape types, each describing an area of the landscape with a distinctive character, visual unity and identifiable sense of place. However, these types should not be considered in isolation. The context of a landscape is an important influence on its character. For instance, views from the crest of the chalk escarpment encompass a whole range of landscape types, interwoven to form a fascinating panorama. By contrast, landscapes such as the narrow river floodplain of the western Rother or the heathlands are visually contained. Many of the landscape types have contrasting scales and patterns and their identification helps us to analyse how the landscape is put together and why it looks the way it does.

This analysis of landscape types also helps to create policies for managing the AONB landscape which reflect the many influences contributing to its distinctive character and local sense of place, so ensuring that these special qualities are conserved.

Aerial photograph of Cuckmere Valley

Landscape Character Areas

The landscape of the AONB can be divided into three broad categories: chalk landscapes, wealden landscapes and river floodplain landscapes. Within this broad classification there is a range of different landscape types, illustrated on the Landscape Character Map below.

River floodplain landscapes

The largest rivers within the AONB are the Arun, Adur, Ouse and Cuckmere which flow southwards from the Weald towards the coast. These rivers have carved down through the chalk plateau to form broad, fairly deep valleys (see *Principal chalk valleys*). Their floodplains provide a strikingly different landscape character which contrasts with the exposed rolling qualities of the surrounding chalk uplands.

The river floodplain landscapes of the AONB are described under the following landscape types:

- *Brooks pastures*
- *Principal river floodplains*
- *Minor river floodplains*

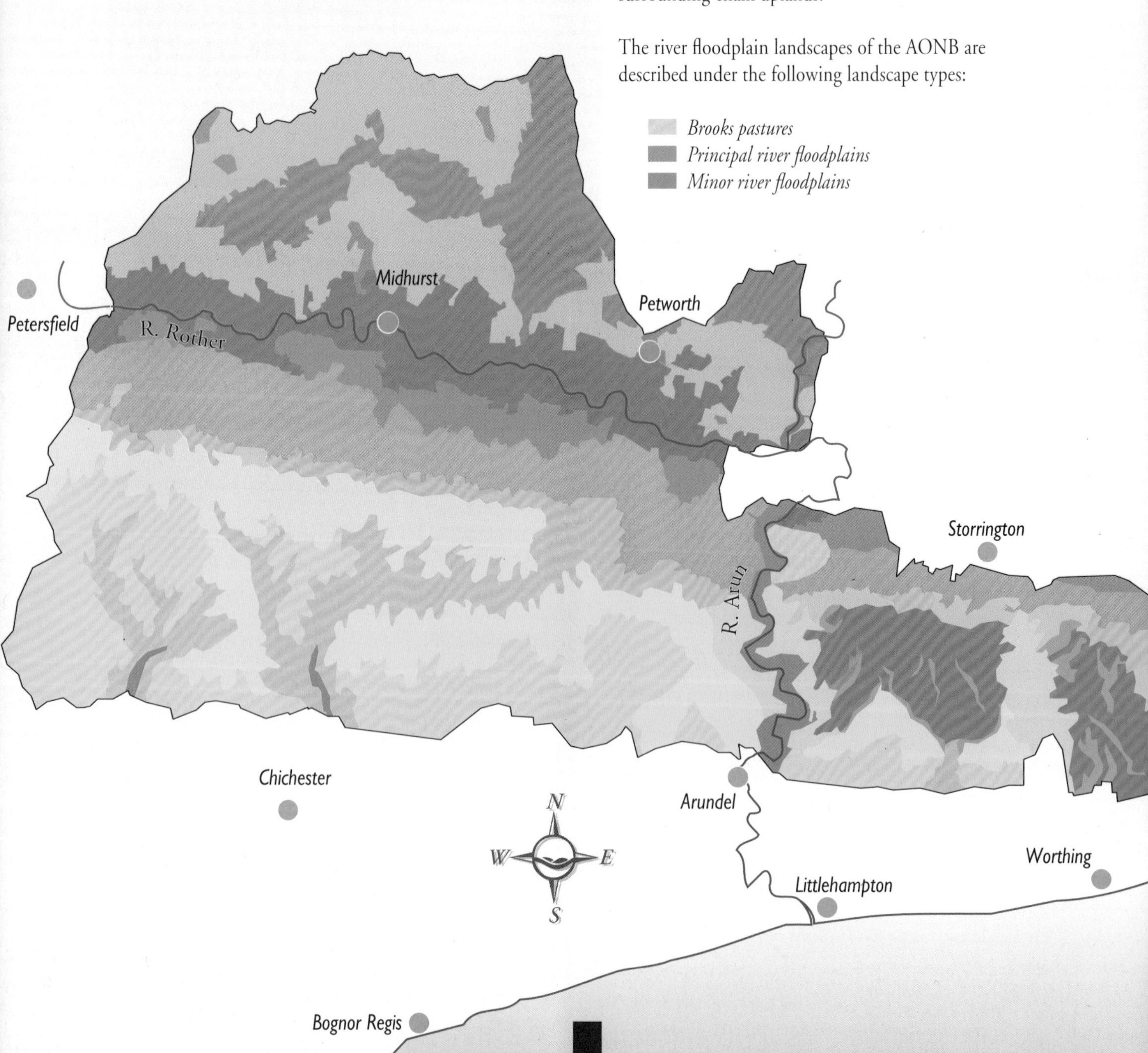

Aerial photograph of Iping and Stedham

Aerial photograph of escarpment at Firle

Wealden landscapes

The Sussex Downs AONB includes the north-west Weald to the west of the Arun valley. This is a complex landscape and variations in landscape character relate closely to the underlying geological structure. To the north of the chalk escarpment, the rocks outcrop in a series of parallel, linear bands of sandstone and clay. The landform dips gently towards the western Rother before rising steeply to the wooded Lower Greensand ridges on the north side of the valley.

The wealden landscapes of the AONB can be divided into the following landscape types:

- *Scarp footslopes*
- *Sandy arable farmland*
- *Heathland mosaic*
- *North wooded ridges*
- *Low Weald*

Chalk landscapes

The South Downs consist of a chalk dip-slope, inclined to the south, and a dramatic north-facing escarpment. The chalk dip-slope has been eroded into numerous small dry valleys and the entire mass of chalk has been carved into separate blocks by the principal chalk valleys.

The chalk landscapes of the AONB can be divided into the chalk uplands of the south-facing dip-slope, the chalk valleys and the north-facing chalk escarpment. Within these categories there are the following landscape types:

Chalk uplands:
- *Open east chalk uplands*
- *Enclosed west chalk uplands*
  - *Small scale*
  - *Large scale*

Chalk valleys:
- *Principal chalk valleys*
- *East chalk valley systems*
- *West chalk valley systems*

Chalk escarpment:
- *Open chalk escarpment*
- *Wooded chalk escarpment*

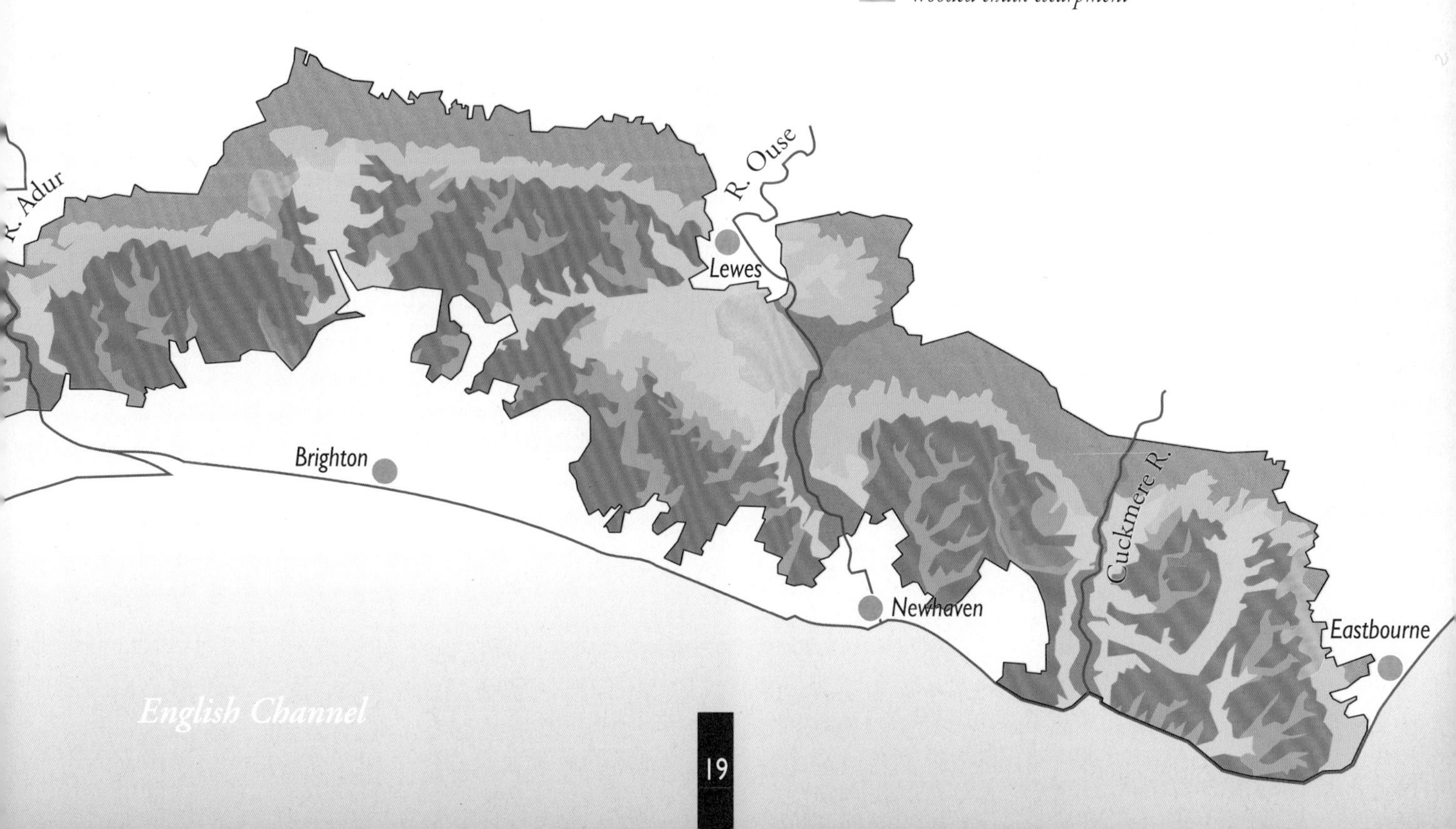

Open east chalk uplands

Key characteristics

Smooth, gently rolling landform with an open, expansive character.

Large rectangular fields; arable farming predominates, with some areas of grassland.

Occasional isolated blocks of woodland and very few hedgerows or hedgerow trees.

Few settlements; only isolated farm buildings, although extensive urban development on coastal plain to the south.

OPEN EAST CHALK UPLANDS

This is an expansive, open landscape of vast panoramas across uniform chalk upland scenery. Uncamouflaged by hedgerows and woodlands, the profile of the broad, rolling chalk ridges stands out dramatically against the sky and dominates all views.

Everywhere the chalk dip-slope has been sculpted into distinctive, sweeping forms by dry valleys. In some areas such valleys are mere indentations, forming a simple rolling relief; elsewhere they are steep, rounded coombes which subdivide the chalk uplands and produce strong contrasts in landscape character. (The latter are described as a separate landscape type: see *East chalk valley systems*.)

Generally the relief becomes more undulating towards the northern and highest part of the dip-slope, but the individual steep rounded summits of the eroded secondary escarpment form a prominent sequence of landmarks, spaced at regular intervals along the southern edge of the escarpment (eg Harrow Hill and Steep Down).

Fields are often extremely large and rectilinear in shape. Intensive arable farming has been the norm since the Second World War, although the visual impact of the South Downs Environmentally Sensitive Area Scheme is increasing and there are now also substantial areas of grassland. Historically, hedgerows have never been a feature of the open east chalk upland landscape. Where they do occur, near farmsteads or alongside ancient chalky tracks, they tend to be narrow and sparse, with occasional stunted trees. Some fields are bounded by wire fences but most remain unenclosed. The arable fields form an extensive, geometric mosaic which varies in colour and texture with different crops and seasonal cycles. The straight lines of the field patterns criss-cross and emphasise the gentle rolling relief.

Small isolated blocks of woodland, usually with strongly regular shapes, are occasional landmark features, visible for miles around. Their dark forms stand out in sharp contrast in a landscape of muted tones. The chalk uplands feel spacious, bleak and very exposed to the elements; skyscapes and weather conditions are a dominant influence, creating a dynamic, moody landscape which can be exhilarating and which feels close to nature, even though it is subdivided by the razor-hard edges of an intensive agricultural mosaic. This is a landscape that can feel remote and wild despite the relative proximity of extensive urban areas and the intensity of agricultural production.

There are very few villages and roads; only occasional farms and isolated barns standing in stark isolation at the end of straight, chalky-white tracks. However, the South Downs is a relatively narrow spine of chalk, hemmed in to the south by the coastal conurbations of Worthing, Brighton and Newhaven. There are few concealed corners in this open, spacious landscape and the roads and buildings of the towns to the south of the AONB are clearly visible, forming an abrupt, harsh boundary to the soft rolling downland. Golf-courses and transmission lines are scattered on the open downland to the north of the towns, in places extending the visual influence of the urban areas far into the AONB.

John Tyler

The Downs from Cissbury Ring

Chalk cliffs

The entire structure of the South Downs is revealed in profile at the range of gleaming white chalk cliffs between Beachy Head and Seaford - a massive chunk of chalk, created over millions of years and sheared off to form a cross-section of geological time.

The retreat from the sea continues today at a rate of almost one metre per year. Ominous white cracks in the cliff-top turf anticipate the next broken edge and the famous Belle-Tout lighthouse, now closed to public access, stands poised at the very edge, awaiting a spectacular end.

Domenic Sweeney

Beachy Head and Lighthouse

The crunchy, skeletal shapes of the microscopic algae which formed the chalk are clearly visible in the sliced off layers of the cliff-face and the pile of broken fragments at its foot. Above, the deeply undulating skyline profile of the Seven Sisters is one of England's best loved coastal landmarks. This sequence of once tranquil dry valleys has been dramatically truncated and exposed, revealing the rolling curves of the chalk landform in profile.

Inland, the extensive greensward of traditional downland, dotted with sheep, reaches right up to the cliff tops, the result of recent land-use policies by Eastbourne Borough Council which has encouraged the conversion of arable fields to pasture at this special finale to the South Downs. Kipling would have approved; indeed he would have been appalled to learn that his favourite piece of cliff-top downland (described in the poem 'Sussex' in 1902) had ever been under the plough:

Clean of officious fence or hedge,
Half wild and wholly tame
The wise turf cloaks the white cliff edge,
As when the Romans came ...

Dew ponds

Chalk is a porous rock; rainwater trickles rapidly down through tiny holes in the rock, leaving a crumbly dry surface. This natural characteristic created problems for watering livestock on the Downs.

Dew ponds are a traditional response. These small, round ponds are lined with puddled clay to retain rainwater and dew formed from condensation. The earliest dew ponds would have been constructed on ridgetops where superficial clay deposits occur naturally, although occasionally clay was carried up on to the Downs to provide an impermeable seal.

Domenic Sweeney

Dew Pond at Ditchling Beacon

Key characteristics

Broad, rolling chalk uplands with a bold, varied character.

Diverse patchwork of open farmland, woodland and commercial forest.

Strong variations in landscape scale; fields and woodlands are larger on the higher ridges and form a smaller, more domestically scaled mosaic on the lower slopes.

Dense hedgerows form an interconnecting network, linking woodlands and copses.

Numerous villages clustered in sheltered sites.

Large estates with great houses and designed parklands.

ENCLOSED WEST CHALK UPLANDS

The combination of farmland, woodland and dense hedgerows superimposed on deeply folded chalk upland relief provides a rich variety of landscape patterns, all with a strong sense of enclosure and a bold, distinctive identity. The landscape of the enclosed west chalk uplands seems all the more diverse and secluded because it contrasts so dramatically with the bleaker, exposed open chalk uplands to the east.

These western chalk uplands have a particularly wide variety of landforms, since some of the valleys have carved wide, deep channels in the chalk, leaving broad, undulating ridges which subdivide the uplands on a larger scale than is typical of the open chalk uplands of East Sussex. These larger valleys contain intermittent chalky streams (see *West chalk valley systems*). Elsewhere, the mosaic of different land uses seems to flow over the undulating slopes of the smaller dry valleys with few abrupt edges or contrasts in character.

The accumulated clay deposits on some of the chalk ridgetops provide soils deep enough to support oak trees and sufficiently acidic for holly and some birch. Beech is dominant on the thinner chalk soils, with yew and whitebeam also common in hedgerows and copses throughout the area.

Woodlands, copses and areas of extensive commercial forestry subdivide farmland in which arable land predominates, but which also includes substantial areas of pasture and smaller fragments of chalk grassland and scrub. The picture is further complicated by the strong influence of several large estates and many designed parkland landscapes, with their distinctive sweeping landforms and carefully grouped clumps of trees. Such estates own substantial areas of the dip-slope. Each encompasses a wide variety of different landscapes and their presence over the years has helped to maintain the typically diverse landscape pattern. Designed features such as avenues, follies and vistas contribute a lively, unpredictable character and give the area a strong local identity. Hundreds of years of management has left many traces, from ancient earthworks to yew forests and stately homes. The landscape has a timeless quality and a strong sense of history.

Areas of higher land tend to have a land-use pattern with a relatively large scale. Here woodland, plantations and fields tend to be extensive, with irregular shapes. These areas are indicated on the landscape character maps as large-scale enclosed west chalk uplands.

There is a very gradual transition to the finer-grain landscape pattern typical of the lower parts of the chalk dip-slope, where fields and woodlands are considerably smaller and form the more coherent patchwork described as the small-scale enclosed west chalk uplands.

Numerous small villages and hamlets are clustered in sheltered sites on the lower slopes. Their traditional flint buildings and walls have a secure, harmonious character. Towards the southern margins of the chalk, buildings are made from a more diverse range of materials, with rendered and brick houses interspersed with those of the traditional flint. Glimpsed through the trees, they provide many contrasts of form and character.

The many villages, distinctive local features and historic parklands give the landscape an idyllic, welcoming character. Much of the area seems deeply rural, but without feeling remote or inaccessible. The rich pattern of fields and woodlands ensures that views are unpredictable and constantly changing. It is easy to lose your bearings and panoramas from highpoints, such as Bow Hill and The Trundle, provide a valuable overview and an opportunity to understand the overall setting and form of the area in relation to its surroundings.

John Tyler

View from the chalk uplands south towards the coastal plain and Chichester Cathedral

Sussex FWAG

Woodland management in the western Downs

Working woodlands

Commercial forestry plantations carpet extensive areas of the west chalk dip-slope, particularly the higher land towards the northern escarpment. Many of these woodlands, such as Selhurst Park and Eartham Wood, are a mixture of broadleaved and coniferous species, and some contain areas of ancient woodland as part of the wooded mosaic.

Traditional parkland estates

This area has many fine stately homes and parklands. These landscaped parks, with their sweeping pastures, ha-has, vistas and carefully placed groups of spreading trees, are special features in the wider landscape, providing an elegant, striking contrast to the surrounding patchwork of fields and woodlands.

Stane Street

The Romans were the first to build engineered roads capable of taking traffic over long distances. One of the best known is Stane Street, which led from the east gate of Chichester directly over the Downs towards Pulborough, Billingshurst and, eventually, London. Pottery finds suggest that the road was constructed almost immediately following the Conquest in AD 43.

Principal chalk valleys

Key characteristics

Broad U-shaped valleys with a fairly flat valley floor and steep, undulating side slopes.

Valley sides are indented with blunt, rounded coombes and deeper, branching side valleys.

Patchwork of arable fields and woodland forms distinctive patterns on valley sides; grassland and irregular patches of scrub are common on steeper slopes.

Settlements usually clustered on the lower slopes above the floodplain, with isolated farms on higher land.

Roads and power lines link villages along the valleys.

PRINCIPAL CHALK VALLEYS

Each of these principal chalk valleys forms a substantial, broad, U-shaped trough through the eroded chalk. The four valleys of the Arun, Adur, Ouse and Cuckmere rivers generally have steep side slopes, often with minor cliffs such as those along the edges of the floodplain at Amberley, whereas the dry valleys containing the villages of Findon, Pyecombe and Jevington have a more gently rounded profile.

The valley side slopes are typically deeply indented by dry valleys, displaying a wide variety of forms and stages of evolution. Some are rounded hollows perched well above the valley floor; others are deeper coombes. There are also many examples of well-developed dry valleys which form deep, winding troughs twisting at an angle away from the principal valley (see *East chalk valley systems*).

Most of the fields on the valley sides are used for arable crops, with permanent pastures on the steeper, more undulating slopes. The steepest slopes, generally at the heads of the minor side valleys, are a patchy mosaic of scrub and rough grazing. Field shapes vary. Most are enclosed by hedgerows, although these tend to be more stunted and sporadic in the exposed arable farmland landscape on the upper valley slopes. Woodlands with irregular shapes often form a distinctive part of the patchwork of land uses. Many have sweeping forms which closely follow contour lines, emphasising the undulating landform along the valley sides.

Tree cover increases towards the valley floor, where the farmland patchwork becomes smaller in scale and hedgerow trees tend to link small woodlands and copses to create a fairly strong wooded edge to the floodplain in the river valleys. Isolated individual trees often contribute to the more diverse landscape pattern on the lower slopes.

Roads and lanes connect the villages along the lower slopes of the valleys, but many act as important major communication routes through the South Downs, linking the coastal urban areas with London's hinterland to the north. Transmission lines and railways also follow these principal chalk valleys. The extensive urban development on the coastal plain has a strong visual influence towards the southern margins of the chalk. Exceptions are the Cuckmere and Jevington valleys to the east, which still have an unspoilt, deeply rural quality. The Cuckmere valley, in particular, is unique in the South Downs as the floodplain remains undeveloped, allowing long unspoilt views along the valley to Cuckmere Haven.

Villages tend to be concentrated in sheltered sites on the lower slopes of the valleys, usually on the very edge of the floodplain. Farms, small groups of farm cottages and isolated barns are often found on the more elevated slopes. Most buildings are partially screened by dense tree cover, but the villages are often a focus within views across and along the valley. Building materials are a diverse mix of flint, brick, timber, rendering and imported stone - a contrast to the more homogeneous character of villages in more isolated parts of the chalklands.

The principal chalk valleys seem settled, domestic and secluded in relation to the windswept open chalk uplands that surround them. Nevertheless, their landscapes have a fairly large scale and views within the valleys reveal the sweeping grandeur of the chalk landform. Most have a strong sense of harmony and order, with land uses responding closely to variations in landform.

John Tyler

The Cuckmere Valley looking south towards the Sea

John Tyler

Chalk cliffs on the River Ouse near Lewes

The Cuckmere elms

The arrival of Dutch elm disease in the 1970s was a particular threat to the valley of the Cuckmere river, which was noted for its magnificent elm trees. Many have been lost, but a local policy to fight the disease's influence is proving an effective defence, if not a total antidote.

The Cuckmere valley lies within a Dutch Elm Disease Control Area (from Brighton to Eastbourne). Aided by the relative isolation and containment of the valley by the surrounding, naturally elm-free chalk uplands, a programme of intensive care has preserved some of the elm population intact. Trees are injected, pruned or felled, and monitored, steadily building up a body of information which, in time, may be used to help restore disease-resistant elms to the English countryside on a wider scale.

Chalk quarries

Chalk quarries often form prominent white scars along the steeper valley side slopes. They are most numerous along the Ouse valley where they seem relatively exposed and, visually, poorly integrated in this rather open valley landscape. The disused cement works in the Adur valley is particularly visually intrusive, not least because the buildings and chimney are so stark and monumental. This quarry has sliced abruptly into the valley side, creating a deep gash which is highly visible in longer views from the surrounding open chalklands. Elsewhere the quarries are less intrusive and many older, well vegetated quarry cliff faces, such as those near Amberley, add interest and detail to views along the valley.

East chalk valley systems

Key characteristics

Relatively narrow, branching valley systems which nevertheless have a strong visual influence and an enclosed, relatively secluded character.

Varied valley forms; some have steep side slopes and lead to deep, narrow rounded coombes; others are asymmetrical in form with very steep, curving slopes on one side and gentle gradients on the other.

Pastures and patchy scrub predominate, with some arable fields on gentle slopes.

Fields often have irregular shapes, following and emphasising the landform.

Individual patches of woodland, hedgerows and hedgerow trees are important features in most valleys.

Villages, farms and hamlets are clustered in sheltered sites on valley floors, linked by lanes and tracks.

EAST CHALK VALLEY SYSTEMS

These branching, linear dry valley systems display a variety of forms and different stages of evolution. The valleys are of two main types: rounded valleys and asymmetrical valleys.

John Tyler

Dry valley

Rounded valleys

The majority of east chalk dip-slope valleys are gently rounded, winding trough-shaped hollows which become progressively deeper and narrower as they cut into the more elevated northern part of the chalk dip-slope. These valley systems usually divide into several branches, each ending in a deep, steep-sided coombe. The landforms seem crisply sculpted and deliberately carved, particularly when the sharp breaks of slope are revealed by shadows in strong sunlight.

The shallower rounded dry valleys are often almost entirely devoted to arable farming, although the presence of a remnant curving hedgerow, fragments of chalk grassland or the odd patch of scrub often indicates a variation in landscape character. The steep side slopes and the rounded heads of the coombes are an irregular patchy mosaic of rough grassland and scrub. Small patches of woodland may also occur and, where they are extensive, provide a particularly strong sense of visual enclosure as in the Stanmer Valley to the north of Brighton. Hedgerows sometimes extend up the valley side slopes and may follow the curving break of slope between the steep valley slopes and the rolling open chalk uplands landscape on the chalk dip-slope.

The sides and floor of the upper dry valleys are often used for pasture but, as the landform becomes shallower towards the south, the valley floor tends to be subdivided into arable fields, leaving progressively narrower fragments of pasture on the side slopes as the valley broadens out on the lower chalk dip-slope. Wherever this occurs, the hard, geometric shapes of arable fields form harsh, ruled edges cutting across the softer, linear grain of the dry valley systems.

These valley systems tend to be too remote and narrow to support large villages but they frequently provide a sheltered site for farms and hamlets which are usually strung out along the narrow lane that follows the valley floor leading to the most distant farm.

The deepest rounded valleys have an intimate scale and a strong sense of enclosure and seclusion. They often seem like a sheltered, hospitable refuge, a quality heightened by the contrast to the relatively bleak, exposed surrounding chalk uplands. The winding, narrow character of these valleys tends to make them seem particularly tranquil, wild and remote even though urban areas and roads may be close by. Many are invisible in long views across the chalk uplands and are therefore surprise features in the landscape, with a special, secretive quality.

Asymmetrical valleys

Curving, up-standing mini-scarp slopes are a prominent feature of some parts of the chalk dip-slope, particularly on the northern slopes of the remnant secondary escarpment hills but also elsewhere on the open downland. The most striking example is Blackstone Bottom to the west of the Cuckmere valley. The formation of these asymmetrical valleys may relate to differences in resistance and therefore elevation of the chalk dip-slope but they may also have formed when the developing valley system eroded at an angle across the natural strike of the chalk.

These asymmetrical dry valleys have a very steep slope on one side (the curving mini-scarp), and a sweeping, shallow slope on the other. The steep mini-scarp slopes are a distinctive, patchy mosaic of woodland, grassland and scrub, forming a striking contrast to the large arable fields on the opposite, gentler side of the valley.

John Tyler

The edge of an asymmetrical valley

John Tyler

Flint building at East Dean

Flint buildings

The isolated groups of traditional farm buildings in these small chalk valleys are built of local flint, often with red-brick dressings edging doors and windows. Farm courtyard walls and barns are made of the same materials, creating a sense of harmony and purpose which makes the groups of buildings seem more substantial and secure than they really are.

Today most traditional groups of farm buildings are dwarfed by their neighbouring modern steel-framed outbuildings, which usually have a universal appearance that denies any sense of place and that reduces the sense of isolation so often associated with the traditional barns.

Field patterns

The slopes of some of the more sheltered chalk valleys have proved too steep and awkward for the plough, preserving the remnants of historical field patterns and villages from as far back as the Iron Age.

Such important archæological sites are a top priority for chalk grassland restoration schemes, with ongoing scrub clearance and carefully monitored grazing regimes. Encroachment by modern farm machinery can destroy valuable evidence in a matter of seconds.

West chalk valley systems

Key characteristics

Broad branching valleys with shallow, rounded slopes and fairly flat valley floors.

Valleys narrow and become progressively more winding towards the north.

Lower valleys have intermittent streams and flat floodplains.

Upper slopes are typically wooded, emphasising the valley landform and enhancing its sense of enclosure.

Arable fields tend to have rectangular shapes, with hedgerows forming straight divisions at regular intervals along the valley sides. Pastures are confined to steeper side coombes and the narrower upper valleys.

Villages linked by winding lanes along the valley floor, with farms on lower side slopes.

WEST CHALK VALLEY SYSTEMS

The broad valleys of the Ems and the Lavant have wide, flattish floors and steep side slopes, strongly indented with rounded coombes. Their defining ridgetops are undulating, but maintain a fairly consistent height and the valleys become progressively narrower as they cut into the more elevated parts of the chalk dip-slope.

The River Ems and the River Lavant are winterbournes, or intermittent streams. For most of the year they are dry and the upper courses of the rivers only begin to flow during exceptionally wet seasons. The rivers only become permanent features to the south of the villages of West Dean and Walderton.

These valleys have a relatively simple landscape pattern. Large, rectangular fields, mainly used for arable crops, stretch across the valley floor and up on to the lower slopes, while the upper slopes and ridgetops defining the valley are generally wooded. There is an abrupt transition to a patchy mosaic of rough grazing and scrub on the steep, curving slopes of the coombes along the valley sides.

Most fields are enclosed by hedgerows which reinforce the strong rectilinear structure of the arable farmland on the floor of the valley, although hedgerows lining the roads along the valley floor have a sinuous form, as do those on the more undulating, steeper valley side-slopes. Here the fields tend to be smaller and more irregular in shape, closely reflecting and emphasising the landform of the coombes. Hedgerow trees are found throughout the farmland patchwork but are concentrated around farms and villages and along the roads. They are usually found in groups or shelterbelts and any isolated individual trees seem conspicuous.

The woodlands on the upper slopes and ridgetops give the valleys a consistent, clear visual structure and provide an overall sense of enclosure. Viewed from within the valley, the skyline is almost continuously wooded. The woodlands are a diverse mix of species - beech, whitebeam, ash, field maple and yew are typical, with oak, birch and holly wherever there are pockets of clay on the chalk. In places small conifer plantations also form part of the mosaic. The yew forest of Kingley Vale is a particularly rare and important feature.

The valleys are well populated with a series of small, clustered flint-stone villages linked by spinal valley roads. The villages occur at regular intervals, but become much smaller in the upper valleys where the river courses are usually dry. Many of the villages are a mixture of houses and farms but there are also farms along the valley roads and in more remote but sheltered sites towards the outer edge of the valley floor. Farm buildings and isolated barns are often a visual focus within this relatively open landscape.

The consistent, repetitive landscape structure gives an overall sense of harmony, order and tranquillity. The valley landscapes become more secluded and enclosed as the landform narrows, although the strong linear, directional quality remains and the changing rhythmic sequences of views along the valleys are one of their most attractive characteristics.

John Tyler

Singleton village

View north from the Trundle

River capture

The Lavant valley is one of only two valleys in the chalk landscapes which has a longitudinal alignment across the dip-slope. This has occurred because the River Lavant, which was originally aligned north-south, captured the westward flowing East Dean stream. It means that this valley has a particularly even profile for much of its course as it has eroded chalk of a fairly consistent elevation.

John Tyler

Yew woodland at Kingley Vale

The ancient yew forest of Kingley Vale

Yew trees were an important resource back in the days of the Hundred Years War, when the longbow was the mainstay for the English archers. Yew was also used for everyday objects which needed a strong, flexible, fairly water-resistant wood such as mill cogs and axles. The tree is renowned for its longevity and some specimens are truly ancient, with twisted, gnarled trunks and dark, spreading branches which extinguish daylight so that the tree stands on an open carpet of fallen berries, needles and twisted broken branches. No plants can survive beneath such a dark canopy.

Today individual trees or small copses are common enough, particularly in churchyards or on historic estates, where they are often the focus for folktales and local mysteries, but at Kingley Vale an entire woodland of ancient yews has survived. It is unique in England, and Kingley Vale is now an important National Nature Reserve, managed by English Nature.

Key characteristics

Steep, distinctive chalk margin with precipitous north-facing slopes and a dramatic undulating ridgeline.

Panoramas from the escarpment ridge with striking views of the Weald to the north.

Predominantly open grassland; patches of scrub and woodland form distinctive patterns; individual landscape elements such as radio masts, chalk pits, clumps of trees and earthworks are prominent landmarks.

Farmland patchwork often extends some way up escarpment slopes, particularly on shallower gradients.

Ancient chalkland track, the South Downs Way, follows ridgeline.

Prehistoric earthworks, in particular Iron Age hillforts, at prominent sites along crest of ridge.

OPEN CHALK ESCARPMENT

The chalk escarpment to the east of the Arun valley generally has an open character, although there are a few small sections of the escarpment further west, such as Treyford Hill, which remain substantially free of woodland.

The open chalk escarpment follows a rather meandering east-west alignment before turning southwards at Polegate to meet the coast at the dramatic white cliffs of Beachy Head. It is deeply indented by steep, rounded coombes and in places forms a series of headlands jutting out into the lowland landscape to the north. Wind gaps occur wherever the principal chalk valleys have broken through the escarpment, providing dramatic opportunities to appreciate the massive bulk of the tilted chalk plateau and to contrast it with the relatively secluded character of the valley landscape. Principal communication routes follow these valleys, so in many respects the escarpment summits can be seen as gateways to the South Downs.

The relatively open character of this escarpment landscape makes the landform of the ridge seem particularly impressive because its strong, undulating profile is uncamouflaged by trees. The escarpment is a series of smooth, rounded summits and ridges, which generally reach an elevation of 200 m, but have high points of 238 m and 248 m at Chanctonbury Hill and Ditchling Beacon, respectively. To the east of Lewes the ridge is marginally lower, around 180 m high.

The northern slopes are consistently steep, and in places precipitous, with abrupt breaks of slope. Deeply indented coombes impart a dynamic, rhythmic quality, particularly when a low sun casts long shadows which ripple across the landform.

From the escarpment summits there are panoramic views over the lowland farmland landscape to the north and also across the chalk dip-slopes to the south. The dry valley systems often end in steep, dramatic coombes, on the dip-slope side of the escarpment. Examples are Well Bottom to the south of Chanctonbury Hill, Hogtrough Bottom to the south of Ditchling Beacon and the Devil's Dyke valley.

The steep escarpment slopes are an irregular, patchy mosaic of scrub, chalk grassland, rough grazing and, occasionally, broadleaved woodland. Hedgerows are infrequent; the field network usually breaks down abruptly as the slopes steepen. The arable fields at the foot of the escarpment often extend some way up the lower slopes, emphasising the undulating line marking the break of slope between the rough-textured steep slopes of the ridge and the managed, relatively small-scale farmland patchwork below.

Mature woodlands with a diverse range of species form irregular strips and patches along the lower escarpment slopes. Stands of beech are found in sheltered locations, and oak stands wherever there are pockets of clay. Other species are ash, field maple, wild cherry and, more importantly, large-leaved lime - an indicator that some of the woodlands are of old semi-natural ecological status. Such woodlands help to 'anchor' the slopes visually and sometimes have distinctive outlines.

Open grassland areas are a muted grey-green colour and are interspersed with patches of scrub, giving the slopes a rough texture and a wild, natural character. Soil creep often causes a hummocky surface and the steepest slopes are wrinkled into tiny terracettes, with rough sheep tracks and footpaths zigzagging along the contours.

The escarpment itself is a landmark, but there are also many particularly distinctive features, with both a positive and negative visual influence. The South Downs Way, an ancient track following the crest of the ridge, often gleams as a chalky-white track as it ascends a grassland summit, and prehistoric earthworks such as Cissbury Ring and the hill-fort at Devil's Dyke are evocative landmarks. But there are also modern, less attractive elements, such as the Devil's Dyke pub and the profusion of masts and buildings on Truleigh Hill. The white scars from abandoned chalk pits are recurring features and some are sufficiently large to disrupt the slope profile of the escarpment.

This is a landscape of dramatic, evocative contrasts. The windswept ridgetop is juxtaposed with the hidden, secretive landscape of the deep coombes on the chalk plateau to the south and spectacular panoramic views across the patchwork maze of the farmland to the north. The ridge always has a strong, enclosing presence but is also a commanding landmark. The massive, smooth rounded summits have a primitive, monolithic quality and there is an overwhelming sense of the power of natural forces.

John Tyler

Chanctonbury Ring

Chanctonbury Ring

In 1760, Charles Goring, heir to the Wiston estate, responded to the then fashionable trend for 'landscape embellishment' by planting a circle of beech trees on the crest of Chanctonbury Hill. Today Chanctonbury Ring is probably the most familiar landmark on the skyline of the chalk escarpment, visible for miles around and a popular destination for walkers along the South Downs Way.

Sadly the finest specimen trees met a sudden end in the storms of the night of 16 October 1987, when the great gale devastated the Ring. New trees have been planted and are thriving, though it will be some time before the smooth, circular profile of Chanctonbury Ring is restored once again.

John Tyler

Devil's Dyke

Dominic Sweeney

Walking the South Downs Way at Ditchling Beacon

Devil's Dyke

This is a curving, knife-like dry valley, incised deep into the escarpment above the village of Poynings and forming a natural outer defensive ditch to the ramparts of the Iron Age hill fort on the summit above.

It has long been one of the most popular 'honey pots' for visitors, who have flocked to enjoy the panoramic views since Victorian times, when a small branch-line took day-trippers from Brighton to the summit and there was a small funicular railway running up and down the steep scarp.

Today the summit is graced by a pub and car-park and there is a golf-course on the ridgetop just to the west. Many come to watch the hang-gliders floating off the edge of the scarp.

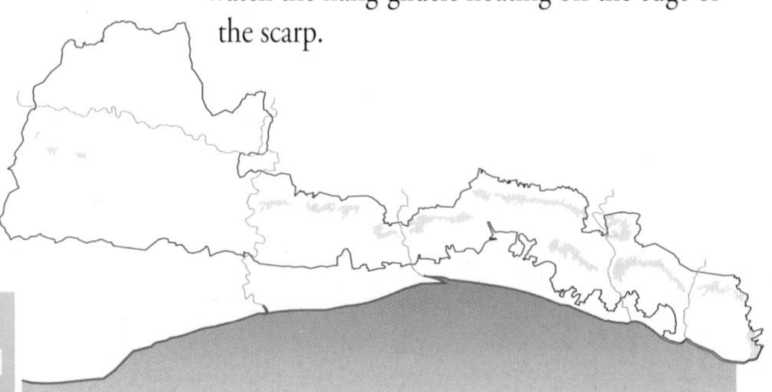

Wooded chalk escarpment

Key characteristics

Steep, distinctive chalk margin with precipitous north-facing slopes and a dramatic undulating ridgeline.

Mixed woodland almost entirely clothes the steep north-facing escarpment slopes.

Occasional patches of open grassland or scrub form highly distinctive landmarks in views from lowlands to the north.

Dense woodland tends to mask variations in relief and emphasises the linear continuity of the ridge as a whole.

Rounded summits are generally open grassland or arable fields but woodland typically extends to the skyline in lowland views towards the scarp.

WOODED CHALK ESCARPMENT

To the west of the Arun valley, most of the chalk escarpment is densely clothed in woodland. In addition, there are wooded sections further east. The largest is on the western fringes of Eastbourne, but there are also examples at Newtimber Hill, the north-east slopes of Chanctonbury Hill and the scarp slopes near Storrington.

This western part of the escarpment ridge remains unbroken by wind gaps, although the relatively low points at Cocking and Duncton are given emphasis by their use as communication routes. The ridge follows a relatively straight east-west alignment which makes it seem particularly consistent. Only at Torberry Hill is there much variation. Here an outlying hill is connected to the escarpment by a ridge, forming a deep, secluded bay in the escarpment to the west of the village of South Harting. Predictably, it was chosen as a perfect site for an Iron Age hill-fort.

The woodlands on the steep scarp slopes are a rich mosaic of beech, yew, field maple, ash and sometimes holly. There are almost pure stands of beech on more stable soils and a mixture of beech and yew on steeper, drier slopes. Many are ancient semi-natural woodlands. Hawthorn scrub predominates wherever there are gaps in the woodland cover. Some of the woodlands on the steep north-facing slopes are a mixture of conifers and broadleaved species, but there are no pure stands of conifers. The beech and yew hangers on some of the most precipitous slopes are of particular visual, historical and ecological interest. The finest examples are between Duncton Down and Bignor Hill.

Areas of open grassland are rare, but where they do occur they tend to be prominent landmarks. The wooded slopes generally have a soft, irregular texture but there are often strong geometric edges on the lower slopes, marking an abrupt transition to the arable farmland landscape at the foot of the scarp. The ridgeline is predominantly open - either grazed grassland or arable fields. However, the woodland on the steep northern slopes extends right up to the skyline in views from the lowlands to the north, and so the open ridgetop landscape is often hidden from view. These exposed arable fields are particularly visible where the ridge undulates steeply at Cocking Down and Manorfarm Down. The South Downs Way is a prominent feature in the long views along the crest of the ridgeline and often acts as a straight-edged division between the large fields.

Mark Brookes

Deciduous woodland at Bepton Down

John Tyler

Deciduous woodland

As in the open chalk escarpment landscape, transmission masts and quarries are prominent landmarks. Examples are the masts on the summit of Burton Down and the quarry on Manorfarm Down. However, the dense woodland cover gives this escarpment landscape more visual continuity than that to the east, and smaller individual features and details have less visual impact.

Viewed from the lowlands to the north, the escarpment resembles a dark, solid, wooded wall. The woodland cover tends to diminish the height of the landform but at the same time increases the impression of continuity and solidity. The scale of the undulating ridgetop relief is best appreciated from the South Downs Way.

The woodland tends to mask individual features in longer views and so there are many hidden surprises, such as the group of ancient tumuli known as the Devil's Jumps on Treyford Hill. The long climb up through dense woodland builds a sense of anticipation for the views from the open summits.

Beech hangers

Some of the most precipitous scarp slopes are clad in 'hanging' woodlands. Most are magnificent beech trees, clinging to the slopes with tortuous, twisted roots, but some are stands of yew, possibly hundreds of years old. Beech trees are particularly susceptible to windthrow and the 'hangers' tend to survive in relatively sheltered sites, with the added protection of the surrounding woodlands. Yews fare better in the wind: they can gain a stronger foothold on the thin, chalky soils and can cope on even steeper slopes. These are ancient semi-natural woodlands of exceptional ecological, visual and historical value.

Tortuous bostels

Known locally as bostels, the deeply sunken, wooded tracks and lanes which climb the escarpment in tight zigzags may be of prehistoric origin. Over the centuries people have used the same established routes to climb the steep scarp. In medieval times, these tracks were the drove roads used by farmers to move their livestock from the Weald to the chalk downland for summer grazing.

Scarp footslopes

Key characteristics

Rolling lowland landscape, dominated by the chalk scarp immediately to the south.

Numerous streams flow northwards from springs at the foot of the chalk escarpment.

Varied patchwork of farmland and woodland interlaced with hedgerows.

Both pastures and arable land; fields have irregular shapes and sizes.

Numerous villages clustered at regular intervals beside springs and millponds.

Dense twisting network of narrow lanes, often deeply sunken between high hedgebanks.

SCARP FOOTSLOPES

The landscape of the scarp footslopes lies immediately to the north of the chalk escarpment. As ever, variations in geological structure are closely reflected in landscape character and landforms in this area are derived from two different types of rock: the Upper Greensand bench at the foot of the chalk escarpment; and Gault, a heavy silty clay which outcrops further to the north.

The width of the Upper Greensand bench varies from approximately 1 km, near the western boundary of the AONB, to a much narrower band to the east of the Arun Valley. Eastwards from the Adur valley it has no visual influence and here the shallow lowlands of the Gault abut the chalk escarpment slopes. The relief of the scarp footslopes landscape therefore becomes less complex and generally flatter towards the east.

The Upper Greensand bench is incised by numerous streams flowing northwards from the spring line near the foot of the escarpment. These streams usually flow in steep, narrow valleys; springs, artificially dammed mill ponds and ornamental ponds are recurring features. To the north of the Upper Greensand bench, the Gault forms a shallow, gently rolling lowland clay vale. Here the streams flow in shallow valleys with many side branches. Those originating at the foot of the chalk escarpment are joined by others arising from springs along the boundary between the Upper Greensand and the Gault. All these minor watercourses flow broadly northwards towards the western Rother.

Throughout the scarp footslopes landscape, the chalk escarpment forms a dominant ridge enclosing the lowlands along their southern margin. The strong contrast between the two landscapes inevitably makes the lowlands seem diminished and intricate in scale when set against the bold, cliff-like ridge.

Panoramic views to the north from the escarpment ridge at Harting Down reveal a fairly small-scale and complex agricultural landscape with a regular patchwork of medium sized, straight-edged fields enclosed by hedgerows. This pattern is broken at regular intervals by sinuous linear woodlands and, further to the north, small blocks of woodlands. Villages, hamlets and farms form small clusters throughout. This is a simplified picture but it is typical of the landscape pattern throughout the scarp footslopes.

The western Upper Greensand landscape consists of fairly large, straight-edged arable fields, but with a smaller-scale, more angular pattern of mixed farming and narrow linear woodlands near the many villages and streams. There are sudden transitions from the relatively open farmland to the more enclosed and intimate landscape of the stream valleys. Narrow, twisting lanes - which may in places be deeply sunken between steep hedgebanks - link the many scattered farms and clustered villages. The farmland throughout is fully enclosed by well-developed hedgerows with many hedgerow trees. The hedgerows link closely with the woodlands, forming an interlocking network which gives the landscape a strongly structured framework.

To the north of the Upper Greensand bench, the pattern of the field patchwork within the clay vale remains largely unchanged, although numerous streams tend to interrupt the layout, making the fields seem less regular in shape. Again the scale of the farmland varies, in particular becoming smaller and less regular around the villages. Arable fields still predominate overall, but there is an increase in the proportion of pasture. The hedgerow network continues across the geological boundary and ensures that the transitions are subtle and in many places blurred.

However, the woodland pattern varies distinctly between the Upper Greensand and clay vale landscapes. On the former the woodlands are generally confined to the steep slopes of the narrow stream valleys, whereas on the clay vale they occur in larger blocks, usually of a similar size and shape to the fields.

Hedgerow trees, in particular individual mature oaks, become visually more distinctive in the clay vale. The lanes seem more open here, with verges, ditches and glimpsed views to farmland through field gates. There is a general increase in tree cover towards the edges of settlements where hedgerow trees link visually with small copses, streamside woodlands and garden trees.

There are far more historic parkland landscapes within the scarp footslopes than in any other part of the AONB. They impose another layer of complexity on the landscape pattern, adding to its diversity and character wherever they occur. Each is an individually designed landscape but they share the characteristic features of the nineteenth century fashion for the picturesque: sculpted, undulating landform, carefully sited groups of specimen trees, lakes, follies and designed 'prospects'. In addition to these larger parklands there are numerous smaller manor houses, farms and mills, all with a contrived, generally secluded landscape setting and with a distinctive character.

This is a well-populated landscape with numerous villages, farms and hamlets reflecting the widespread availability of water from local wells. Most of the settlements are small, but those to the east of the Arun tend to have more suburban influences. The villages vary in form. Some, such as Bignor and Poynings, are fairly clustered while others, such as Graffham and Fulking, have developed in a linear pattern. Most contain farms and display a strongly hierarchical collection of buildings centred on a manor house, church and often a mill. The influence of the larger country houses and their estates is strong in some settlements.

Building materials are very varied: a mixture of flint, brick, a yellowish sandstone, rendering and half-timber. A local chalk rock known as 'clunch', with a washed-out, whitish colour, is widely used as a building material in the Harting area. The varied use of materials gives these villages an interesting, unpretentious character.

The numerous villages, scattered farms and designed parklands give the area a welcoming lively character, with a wealth of detail and interest. Views are fairly contained and many of these features are partially hidden, making this seem quite a secretive landscape which invites exploration and which has many surprises. It is only from the escarpment that the overall pattern of the landscape is fully revealed.

Julian Gray

Burton Mill Pond

Lakes and millponds

In many instances the footslopes streams have been dammed to create mill ponds and the artificial lakes associated with designed nineteenth century parkland landscapes. The most impressive example is Burton Mill Pond at Burton Park, but there are smaller lakes at Parham Park, Wiston Park, Firle Place and Glynde Place, to name but a few.

The underhill lane

The 'underhill lane', the remnant of historical routes, runs along the foot of the escarpment. It is sometimes a road or a lane, but often occurs as a track or bridleway. It is usually deeply sunken between steep hedgebanks. Such sunken lanes, like the narrow stream valleys, are hidden, secretive landscapes, in strong contrast to the medium-scaled, regular patchwork of farmland they cross.

The various sections of the underhill lane probably date from medieval times, when roads were muddy links between villages. There was no overall plan, but the geography of settlement in the area, with a relatively linear sequence of springline villages along the foot of the scarp, meant that a continuous if rather twisted route developed.

Sandy arable farmland

Key characteristics

Rolling relief; the western Rother flows in a relatively wide, shallow valley.

Large arable fields with bold, geometric shapes.

Deeply sunken, straight narrow lanes between steep bracken-covered banks.

Individual, isolated over-mature hedgerow oak trees.

Open, fairly large-scale landscape, although views are often constrained by high hedgebanks.

Small clustered sandstone villages; scattered pattern of farms and farm cottages.

SANDY ARABLE FARMLAND

The sandy arable farmland landscape is confined to areas underlain by rocks of the aptly named Sandgate Beds. This is an easily eroded sandstone which outcrops to the north of the scarp footslopes and heathland mosaic landscapes in the north-west section of the AONB.

This landscape is drained by the western Rother, which flows in a fairly wide, shallow valley from west to east to join the River Arun near Pulborough. The rolling farmland slopes gently towards the river and the area is bounded to both north and south by slightly higher well-wooded land.

This landscape is characterised by a fairly simple visual and land use structure. A regular and relatively large-scale patchwork of rectangular arable fields forms a broad grid across the rolling landform. The fields are generally enclosed by hedgerows, although there is considerable evidence of field enlargement and hedgerow removal. Many of the hedgerows have over-mature specimen oak trees. Small woodlands often link with the hedgerow network and tend to provide shelter and a softer landscape setting near farms.

Narrow lanes, running in a strict north-south alignment across the Rother valley, give access to isolated groups of farm buildings. All routes tend to follow a straight alignment, but with definite kinks, responding to and accentuating the rather regimented pattern of the fields.

The farmland landscape has a fairly open character, although views from the narrow lanes are constrained by very steep, high bracken-clad banks, sometimes contained by sandstone walls. The lanes have become deeply sunken as the soft sandstone has eroded away and they form hidden narrow corridors through the farmland. In places they almost seem like tunnels. This patchwork of arable fields forms a relatively narrow open band between the more densely wooded, finely grained landscapes of the heathland mosaic and north wooded ridges. Its open character provides a sense of scale in the wider landscape and allows some welcome longer views across the Rother Valley, particularly from the higher ground to the north.

The historic parkland of Cowdray Park has created a more wooded landscape to the north-east of Midhurst. The carefully placed groups of specimen parkland trees are a striking contrast to the surrounding geometric farmland layout and those on the rising ground to the north of Benbow Pond are a prominent feature in views across the valley from the south.

Villages are generally clustered at road junctions. There are numerous small hamlets and many scattered farms and cottages. The small market towns of Petworth and Midhurst are on the edge of the area and it is traversed by an important east-west route, the A272, making it the most accessible part of the north-west AONB.

John Tyler

Arable farmland in the Weald

John Tyler

A view southwards towards the Downs

Sandstone estate villages

Many of the villages show the strong unifying influence of the estates of Petworth and Cowdray, with distinctive, well-detailed estate cottages and stone walls. Buildings owned by the Cowdray estate are immediately identifiable by the bright yellow paint on their windows and doors.

The local warmly coloured sandstone is favoured as a building material and varies from a mottled grey or buff tone near Petworth to a much deeper reddish-yellow version in the Midhurst area. There are also many brick and rendered buildings, particularly near Midhurst.

John Tyler

Estate cottages

Hedgerow oak trees

In many areas there are relatively few hedgerow trees, but occasional isolated ancient hedgerow oak trees are important distinctive features of the local landscape. Many are stag-headed and most are over-mature. They have often been left in areas where the surrounding hedgerow network has been removed. Many have clearly reached the end of their life, no doubt hastened by mechanical damage and compaction caused by modern farm machinery. It would be a shame to let such a distinctive local landscape feature disappear for good and it may be appropriate to plant a new phase of specimen oaks, individually spaced so that each tree can eventually fill out to form a fully rounded crown.

Heathland mosaic

Key characteristics

Irregular patchy landscape mosaic of predominantly oak-birch woodland, with conifer plantations, open heath and common land.

Open areas are generally rough grazing, with extensive areas of bracken scrub and occasionally heathlands where heather species predominate.

Dense regenerating woodland thickets surround most open areas and are actively encroaching on to adjacent land.

Few roads; generally straight, often dead-end lanes and tracks lead on to heathland from small settlements on its periphery.

Ancient earthworks and banks.

HEATHLAND MOSAIC

The heathland mosaic landscape occurs on outcrops of the Folkestone Beds and the Hythe Beds, both types of Lower Greensand. It is most extensive on the Folkestone Beds to the south of Midhurst where a broad swathe of heathland on fairly low flat-topped ridges stretches from Iping Common to Duncton Common. Further north, on the dip-slope of the Hythe Beds, more isolated patches of heathland are found within the woodland glades of the north wooded ridges of the western Weald. Examples are Weaver's Down, Chapel Common, Woolbeding Common and Hesworth Common.

The heathland mosaic is a patchy mosaic of oak-birch woodland, conifer plantations, open heathland, acidic grassland, bracken and rough agricultural land. These components vary in size and scale and are in a constant state of flux as a result of management practices and natural regeneration.

Tracts of pure heather heathland occur only on sites such as Iping Common and Ambersham Common, which are actively managed to conserve an open heathland habitat. The predominant land use is pine forest and mixed plantations, with extensive areas of oak-birch woodland bordering most open areas. Open patches of bracken and rough grazing land, fringed by scattered birch trees and dense regenerating woodland, are typical. The underlying sandstones have proved a rich source of sand for the construction industry and the heathland mosaic has been extensively quarried. There are many disused pits, as well as a few which are still active, although all are screened by dense woodlands.

There are very few hedgerows and much of the landscape appears to be in transition, with little structure or predictable order. The conifer plantations and straight roads give some hard edges but the visual structure of other areas is governed by natural rather than man-made forces.

Historically the heathland mosaic has been a marginal landscape for settlements and even today there are very few villages, only isolated cottages, farms and some more recent suburban developments. A high proportion of the land is common land, traditionally used by the poor for rough grazing and bracken or turf-cutting. This traditional land use is reflected by the extensive network of tracks and footpaths and by the many long-established villages on the outer fringes of the heathland.

The heathland mosaic seems an untamed landscape. It is well enclosed and sheltered, yet nature seems very close and the dynamics of the mosaic are structured by the natural forces of regeneration. The rich profusion of close-knit, tangled natural forms and visual containment can evoke feelings of vulnerability, but this is also an intriguing, mysterious landscape with a wealth of hidden features and detail. The numerous paths and tracks and the traditional use of heathland as common land help to make it a relatively accessible wilderness.

Gill Callander

Iping Common

Ancient Bronze Age barrows

The heathlands were cleared and well settled during the Mesolithic period, when the light woodlands growing on the sandy, impoverished soils would have been relatively easy to penetrate. The early settlers began the process of burning, clearance and stock grazing, albeit on a limited scale, which was continued by later civilisations.

The heathland mosaic landscape is rich in archæological sites. The most significant are the Bronze Age barrows which are often found on higher land. Such features seem all the more intriguing as they are well-hidden secrets within the wooded mosaic.

Julian Gray

Birch invading lowland heath

Julian Gray

Iping Common

Julian Gray

Erosion on heathland

Lullington Heath

Chalk heathland seems an anomaly. How can a habitat associated with plants that are adapted to acidic conditions occur on chalk, a form of limestone? It is possible, but very rare, and Lullington Heath is the best and most extensive example in the country. Chalk heathland has a wholly different landscape character from the enclosed acid heathlands of the Lower Greensand. Here the heathland mosaic forms a finely grained pattern of heather, scrub and small patches of woodland on an exposed chalk upland area. This relatively unenclosed heathland is prominent in long views across the chalk uplands and forms a sharp contrast to the smooth, managed arable fields which surround it.

North wooded ridges

Key characteristics

Prominent high plateau of Lower Greensand, with a steep horseshoe-shaped escarpment enclosing the Milland Basin.

Varied, undulating landform; streams flow in deep gullies.

Dense, mixed woodland predominates, with pastures and rough grazing within clearings and beyond woodland edges.

Diverse woodland structure: overlapping mosaic of broadleaved species, conifers, commercial plantations and coppice.

Narrow twisting lanes, often deeply sunken, link isolated farmsteads and local suburban development.

Deeply enclosed landscape with a strong sense of mystery.

NORTH WOODED RIDGES

The north wooded ridges landscape is found on the relatively resistant Lower Greensand rocks of north-west Sussex. The land rises steadily to the north of the Rother Valley and then drops away abruptly at a steep, deeply indented escarpment. The scarp curves around to the north, enclosing the Milland Basin, before twisting northwards to form a series of sharp ridges, separated by deep linear valleys to the south of Haslemere. One such ridge rises to a height of 280 m at Black Down, the highest point in Sussex and an important regional landmark.

Landslips in some areas have led to a stepped escarpment profile, with hummocky relief at its foot. The Lower Greensand dip-slope has an undulating landform, carved into deep ravines, or ghylls, by fast-flowing streams. This complex relief is completely masked by dense woodland which extends across the scarp slope and down into the clay vale at the foot of the ridge.

The forest is a diverse and patchy interlocking mosaic of different species and woodland structures. Extensive conifer plantations and mixed woodlands are interspersed with mixed broadleaved woodlands - many of which are classified as ancient semi-natural woodlands. They are mostly an oak-birch-chestnut mix but also include impressive stands of beech and areas of chestnut and hazel coppice.

Within the woodland, the tree cover is less dense than it seems from the outside. Clearings of different shapes and sizes break up the canopy. Most are small rough paddocks with fairly acidic grassland, usually used for grazing horses. These small fields seem unstructured, with few hedgerows. Most are bounded by wire fencing and have the character of smallholdings, with isolated groups of foresters' cottages clustered on their edge. All are completely enclosed by dense woodland cover and often have irregular indented shapes. Elsewhere open glades contain patches of dense undergrowth and regenerating birch trees. More substantial, hard-edged open areas are found on the crest of Telegraph Hill, where there are some large arable fields, and within the commercial softwood plantations.

Some woodland clearings, such as Chapel Common and Woolbeding Common, are open heathlands (see *Heathland mosaic*) and it is likely that there would be the potential to develop heathland vegetation elsewhere on these poor sandy soils if the tree cover were cleared and the appropriate seed sources available. Small ponds and marshy areas are a recurring feature along the foot of the greensand escarpment. They are often hammerponds, dammed in the sixteenth century to provide a steady flow of water for the forges of the wealden iron industry.

On the clay vale, beyond the foot of the greensand scarp, the woodland clearings tend to have a more structured character, with a network of hedgerows, graced with spreading oak trees, subdividing a patchwork of small fields. Historically, the settlement pattern of this area is one of scattered, isolated farms and smallholdings in the forest clearings and small, clustered spring-line villages, such as Henley and Bexleyhill along the foot of the escarpment. This traditional settlement pattern is intact, although recent suburban development, particularly on the southern fringes of Haslemere, has a strong local influence.

The twisting sunken lanes, hidden clearings, sudden precipitous slopes and ancient gnarled trees all help to create a very strong sense of mystery. This is a secretive landscape which feels hidden, separate and enclosed. The landscape changes rapidly, from coppiced chestnut to stands of pine and ancient oaks, and from undulating pastures to swathes of heathland and glades with marshy ponds. There are many contrasts and surprises; these woodlands are a disorientating maze of lanes, tracks and narrow sunken paths and seem undiscovered, remote and very special.

Traditional woodland management

The diverse mosaic of different woodlands stems from a long history of woodland management. These sandy soils yielded a poor return as farmland, but would have been highly valued as forest. For centuries its trees were a vital resource. Wood was in constant and growing demand for building construction and for making an enormous range of domestic implements, as well as for fuel, charcoal and to supply the local iron and glass industries.

A continuous supply of wood was ensured by a range of management techniques. Of these, coppicing was the most widespread and areas of chestnut and hazel coppice are still an important component of the woodland today. Coppicing is a way of prolonging the life of a woodland, as young sapling shoots constantly resprout from the basal stools after they are cut. Oak trees were often allowed to grow up amidst the hazel coppice until they were ready to fell for timber, a type of woodland known as coppice-with-standards. Elsewhere parts of the woodland would have been grazed as wood-pasture, to provide additional income for the local foresters.

John Tyler

Chestnut stakes

Julian Gray

Chestnut coppice

Wooded ghyll valleys

The greensand ridges are dissected by deep ravines, where the sandstone glistens with moisture and there is a constant gurgle of running water. The ancient woodland tracks and lanes are often also sunken between deep banks and jagged outcrops of sandstone. The lanes near Chithurst are especially deep and the dappled sunlight of the surrounding woodlands is extinguished as they tunnel through the rock.

Low Weald

Key characteristics

Gently undulating lowland vale; numerous small streams, which occasionally flow in steep, narrow gullies.

Relatively small-scale patchwork of pastures and deciduous woodlands interspersed with larger arable fields towards edges of the vale.

Woodlands tend to be strong linear features between fields, often following winding streams.

Many woodlands are ancient and particularly species-rich.

Dense hedgerows containing mature hedgerow oaks enclose fields and link larger woodlands.

Numerous villages clustered around village greens; scattered farms and cottages linked by a network of narrow, winding lanes.

Scenic, deeply rural, tranquil landscape with a domestic character.

LOW WEALD

This is the western periphery of the extensive geographical region known as the Low Weald which extends across north-east Sussex. It has a mixed geology. Dense Weald Clays are interspersed with thin bands of more resistant limestones and sandstones, forming gently undulating relief. The north wooded ridges shelter these lowlands, providing a strong, dark backdrop to local views. The Low Weald is drained by numerous branching streams; many have carved into the Weald Clay and flow in very steep, narrow valleys.

Deciduous woodland, pastures, streams and hedgerows are closely interwoven to form a finely grained landscape pattern with an irregular, organic character. This is predominantly a dairy farming area, although there are some arable fields. The pastures are enclosed by a well-developed hedgerow network. Single mature oak trees are a feature of most hedgerows and are sometimes found as free-standing specimens within pastures.

Many of the woodlands are on the slopes of narrow stream valleys and have a linear, sinuous form. The shape of these 'shaws' tends to enhance the overall sense of enclosure. Historically, woodlands in this area were intensively managed for fuel and timber by a combination of coppicing, thinning and wood pasture. Many are ancient semi-natural woodlands, containing a particularly rich diversity of species. The predominant species is oak, but birch, holly, beech and yew are all represented. The shaws act as important 'green corridors', interconnecting other larger woodlands and harbouring a diverse flora and fauna.

Ponds, marshes and damp, low-lying meadows are, like the streams, well hidden amongst dense tree cover. Some larger ponds, such as that at Mill Farm, are hammerponds dating from the days when the area was dominated by the wealden iron industry. Later they were used as mill ponds, for grinding corn.

John Tyler

Sunken wooded lane in the Low Weald

The larger, expanded villages such as Fernhurst have a rather suburban character, but elsewhere the historical settlement pattern of small, picturesque clustered villages and numerous scattered farms and cottages remains intact. Many, such as Lurgashall, are centred around large village greens. They are connected by narrow twisting lanes.

Over the years, a wide range of local and imported materials has been used. Red brick, sandstone, half-timbered, tile-hung and rendered buildings are all represented. Groups of farm buildings often form an attractive focus in local views, always partially hidden by trees.

The Low Weald has a domestic, deeply rural character. Small villages clustered around greens, commons, ancient hedgerow trees and the landscape's irregular, organic pattern of pastures and shaws all contribute to the almost medieval character of the area. The twisting lanes, irregular small fields and sinuous woodland patterns form a dense maze in which distances seem extended and all destinations seem remote. The profiles of the surrounding wooded ridges are a constant point of reference and an aid to orientation.

There are many shades of green in summer and a profusion of textures and natural forms. The spreading silhouettes of ancient oak trees, deep ferny ravines, undulating damp pastures and winding hedgerows all add to the layers of detail. There are no abrupt edges or sharp contrasts and the landscape has a slow, relaxed pace with a strong, calm sense of security.

Hedgerow trees at Lurgashall

Ancient shaws

Shaws are long, narrow belts of woodland which lie along field boundaries throughout the Weald. Within the Low Weald of the AONB, they tend to follow the tiny sinuous streams, often clothing the steep, damp stream-banks, which are relatively inaccessible for farming.

Many shaws are ancient semi-natural woodland, a diverse coppice of oak, ash, hazel, hawthorn and field maple. Yew is also common and there is a rich ground flora of ferns, mosses and wild flowers. The shaws are thought to date from the original clearance of the wildwood, when small fields were hacked out from the forest. It seems likely that they have been preserved as working field boundaries, intensively managed to produce coppiced timber and underwood, as well as providing shelter and security for livestock.

The Mens

An ancient relic of the original wildwood, The Mens is centuries old. This former woodland common, grazed as wood-pasture by the livestock of poor labourers, has been ungrazed since the beginning of the century.

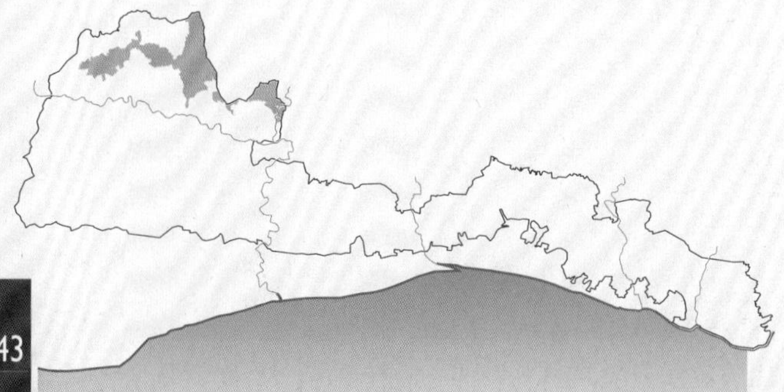

Brooks pastures

BROOKS PASTURES

The brooks pastures are found in the valleys of the River Arun between Pulborough and Arundel, near Amberley, and of the River Ouse, near Lewes. In both instances, a particularly extensive floodplain landscape has developed because the river has been unrestricted by the surrounding landform and has been able to meander in broader loops across a wider area. The floodplains have an uneven surface, with marshy oxbows and water-logged depressions marking abandoned earlier courses of the river.

The fine, silty alluvial clays deposited on the river floodplain form heavy, poorly drained soils, suitable only as pasture land. Traditionally, the floodplain was managed as flood meadows, drained by a geometric grid of narrow channels. These are still intact, though now the flow of water is controlled by sluices. Much of the area is subject to seasonal flooding. The reeds in the ditches provide a striking contrast in texture at the edges of the fields but do not disrupt the open, flat character of the floodplain landscape. The ditch systems and some wet grasslands have a particularly rich flora and attract nationally important populations of winter birds.

John Tyler

Winter flooding

The brooks are mostly used for grazing and for the production of hay and silage. Hedgerows are sparse, but provide a partial network in places. Isolated, stunted hawthorn trees and bushes are scattered throughout the pastures and there are occasional oak and ash trees, together with groups of scrubby willows and alders. The small, isolated woodlands within the Amberley Wild Brooks landscape make this area seem slightly more enclosed and intimate than that of the Lewes Brooks. One of these patches of woodland, at Amberley Swamp, has a very special, mysterious atmosphere. An eroded embankment encircles an area of hummocky marsh with twisted, partially fallen willows and scattered hawthorn scrub to the side of the river embankment. Amberley Swamp seems a forgotten corner of the landscape and its irregular form contrasts with the strict geometry of the surrounding drained pastures.

Farms and villages are often sited on the very edge of the floodplain and are important in local views. Amberley Castle, in particular, is a striking landmark.

The brooks pastures have a domestic, tranquil and expansive atmosphere. Amberley Wild Brooks is deeply rural and seems particularly remote, wild and secretive. There are no visual intrusions and the inaccessibility of much of the area adds to the sense of tranquillity. By contrast, urban development, roads and transmission lines make the Lewes Brooks seem part of this busy, wide open valley. Here the sweeping high chalk ridges form a dramatic backdrop to an extensive and rather unkempt floodplain landscape of rough paddocks, bushes and patches of scrub.

Key characteristics

Flat, extensive alluvial floodplain landscape.

Fairly small, rectangular rough pastures, edged by straight, reed-filled drainage ditches. Many are seasonally flooded meadows.

Relatively open landscape with few hedgerows; occasional patches of woodland and isolated trees.

Deciduous woodlands and dense, well-treed hedgerows on outer margins of floodplain form strong visual edges.

No settlements. Occasional raised straight tracks lead from villages on the margins of the floodplain to established river crossing points.

John Tyler

Amberley Wild Brooks

John Tyler

Sheep grazing by wet fence

Wet fences

Ditches are the 'wet fences' of the brooks pastures, derived from the network of drainage channels dug to drain the waterlogged floodplain when it was first reclaimed to form seasonal flood meadows for grazing flocks of downland sheep.

The ditches are colonised by both stream and marshland species and harbour a wealth of wildlife. Sharp-leaved pondweed and rootless duckweed are just two examples of nationally rare and declining wetland species which are thriving in the ditches of Amberley Wild Brooks. They also conserve an important range of beetles, dragonflies, damselflies and snails. Today, with constant modification of the rivers, the ditches may be a refuge for lowland river species no longer commonly found in the rivers themselves.

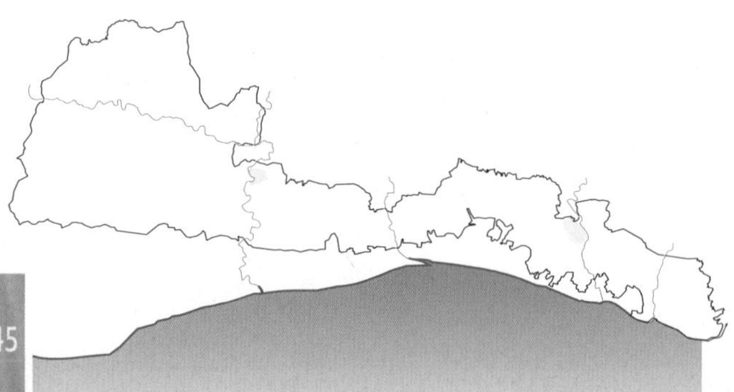

Principal river floodplains

Key characteristics

Flat, open alluvial floodplain landscape of principal river valleys.

Relatively small-scale pastures with rectilinear fields; subject to seasonal flooding.

River typically flows within open, grassed embankments and the floodplain is sliced by drainage channels and canals.

Ponds, reedbeds and marsh edged by patchy willow and alder scrub.

Relatively few trees or hedgerows; many fields enclosed by wire fencing or ditches.

Major roads often mark the boundary between floodplain and valley slopes; railways are typically built within floodplain on embankments.

PRINCIPAL RIVER FLOODPLAINS

The principal river floodplains are found on the flat valley floor of the valleys of the Arun, Adur, Ouse and Cuckmere rivers. The rivers meander across the floodplain in broad loops, which tend to increase in size towards the south. In each case the main channel is tightly enclosed by embankments; there are frequently abandoned isolated remnants of curving embankments elsewhere on the floodplain, indicating former alignments of the river. Many show signs of an artificially straightened course and winding minor tributaries often link with the straight man-made drainage ditches on the floodplain before joining the main river.

The floodplain landscape of the Cuckmere Valley is particularly scenic, reflecting the tranquil, unspoilt character of this valley and its relatively intimate scale. The Cuckmere River is the only river which flows into the sea in the AONB and Cuckmere Haven, at the mouth of the river, is therefore a very special, relatively unspoilt landscape.

Formed on alluvial heavy silts and clays deposited by the river, these floodplain landscapes are predominantly used as pastures for cattle, but there is also some horse grazing and silage production. Arable fields are sometimes found in areas with gravelly river terrace soils towards the edges of the floodplain, at points where the river has tended to deposit rather than erode material.

There is typically a strong contrast between the relatively open, flat pastures of the floodplain and the undulating, well-treed village/mixed farming landscape pattern on the lower valley slopes. The contrast is emphasised by the curving linear strips of woodland which often mark the boundary between the two landscapes. Where these woodlands also screen roads and development, they help to give the floodplain landscape a fairly secluded character.

John Tyler

High and Over

Cuckmere valley

The floodplain pastures are small or medium sized fields of various shapes: some are laid out in planned rectangular plots, while others have irregular wandering boundaries. Most are bounded by narrow drainage channels, which in places break down to form odd abandoned ditches and areas of hummocky marshy ground. Marshy areas and pools of water on the outer edges of the river embankments often have extensive reed beds. The drainage ditches themselves are reedy and often form strong contrasts of texture with the grazed pasture, particularly where there are no hedgerows to interrupt views across the plain.

Hedgerows provide a strong, visual structure in some areas but the network is sporadic and discontinuous. In places it is non-existent and the fields are bounded by wire fencing or left unenclosed and separated only by drainage ditches. There are a few widely scattered small patches of woodland within the floodplain, but the landscape is relatively open and the strips of woodland marking the edge of the floodplain therefore have a strong enclosing influence.

The character of each of the river floodplains varies according to the degree of urban influence. In the quieter valleys the floodplains have a domestic, secluded atmosphere; in the more developed valleys, busy roads and power lines can have a strong influence. Nevertheless, the enclosing valley slopes always give an impressive sense of scale and security and the meandering river provides a rhythmic focus, particularly in views from the adjacent valley ridgetops.

Ports and canals

The Sussex coast is renowned for its shifting shingle spits. Over the centuries the entrances to each of the major ports have shifted eastwards as the harbour mouths became obstructed by the drifting coastline. Further inland, Arundel, Bramber and Lewes are all former ports, long since marooned inland by silt. Their economic importance stemmed from their position at the junction of the two agricultural economies of Sussex: the sheep/corn economy of the Downs and the cattle/timber economy of the Weald.

A spate of canal-building, largely funded by the third Earl of Egremont in the early nineteenth century, led to the cutting of the Arundel and Portsmouth Canal (with its branches to the City of Chichester) and the Wey and Arun Canal. Together these canals and their associated river navigations provided an inland waterway route from London to the sea.

Minor river floodplains

Key characteristics

Narrow, flat alluvial floodplains, closely following the winding river.

Small pastures with irregular shapes and an unkempt character.

Curving, narrow strips of deciduous woodland along river or field boundaries.

Overgrown hedgerows with dense tree cover; occasional ancient hedgerow oaks.

Small farms set back on outer edge of floodplain; hamlets at bridging points.

Narrow humpbacked stone bridges are often important local features along the western Rother.

MINOR RIVER FLOODPLAINS

The minor river floodplains of the AONB are those of the upper Arun to the north of Stopham Bridge, the western Rother, and the lower reaches of both the Lavant and the Ems on the margins of the chalk dip-slope to the north of Chichester.

These are relatively narrow, linear landscape corridors, confined to an alluvial strip along the river. At a local scale, the floodplain is often quite hummocky and uneven, with minor ditches, tributaries and marshy low points. The floodplain of the Rother has many abandoned meanders and, in places, sections of the disused nineteenth century Rother Navigation, which briefly allowed boats to reach Midhurst.

The floodplains of the Rother and the upper Arun are used as pasture for cattle or horses. Fields are generally small and irregular in shape, enclosed by dense overgrown hedgerows. In places the hedgerow pattern has partially broken down and the pastures have an unstructured character with stock wandering freely from one field to the next. Remnants of hedgerows often seem to be freestanding within fields and there are many small, irregularly shaped copses and groups of trees which contribute a sense of randomness to this finely grained, rather unmanaged landscape mosaic.

The curving linear strips of deciduous woodland often mark the outer edges of the floodplain and provide a strong sense of enclosure and seclusion. They are often found directly alongside the river channel itself, or marking an abandoned meander. These woodlands are principally oak, but those near the river channel or marshy parts of the floodplain contain a high proportion of willow and alder. Ancient hedgerow oak trees are prominent local features, particularly along lanes or tracks.

The chalk streams flow within a more ordered landscape of small, flat rectangular pastures. The river channel here has a fairly open character, with little tree cover. Pastures are often edged by ditches as well as hedgerows.

There are no settlements on the floodplain itself, but farms are often set back on the outer edges and there are sometimes small villages or hamlets clustered on the edge of the floodplain at bridging points. There are a number of rights of way, both across and along the floodplain.

This landscape has a strong, distinctive character and presence, despite its narrow form. The intimate scale and organic, random quality of the pastures along the River Rother are in striking

Rother valley

contrast to the surrounding rectangular arable fields. The river floodplain feels like a special, secluded, secretive landscape where natural forces predominate and where human influence is relatively ineffective. The surrounding woodlands and dense hedgerows provide a strong sense of enclosure and the floodplain seems like a separate, tranquil world.

© Nick Meeres

Medieval bridge at Stopham

Medieval humpbacked bridges

The steep old humpbacked stone bridges are an important feature of the landscape along the western Rother. They are made of the local honey-coloured sandstone and are particularly attractive and well-detailed. Their height is an indication of the changeable character of the Rother, which is inclined to surge dramatically. There is evidence of significant flooding in the late medieval period and it seems likely that these over-scaled, beautiful bridges were built in response.

4

Features of the Landscape

John Tyler

The AONB has a wonderful and valuable landscape heritage, the product of centuries of interaction between nature and man. In Sussex, the particular combination of geology, climate and human exploitation has provided us with a wealth of contrasts. Open, rolling chalk grassland, ancient forest, acid heaths, flood meadows, crumbling cliff faces and deep, shady ghylls are all represented within a relatively small area.

Such strong variations produce the beautiful scenery for which Sussex is renowned, but they also have enormous ecological value. Ancient habitats, and in particular those managed in traditional, sustainable ways by farmers and foresters, remain exceptionally rich in wildlife diversity. They have become rarer, and more valuable, as they have been progressively fragmented by the large-scale land use changes required by modern society, particularly agriculture and urban development. This south-eastern corner of the country is extremely vulnerable, making Sussex's riches all the more precious.

Chalk Grassland

The chalk downland is one of Britain's best-loved landscapes. People talk fondly of long, open views, an exhilarating sense of freedom and of the springy, scented, grey-green turf which carpets the hills. Less than a hundred years ago flocks of sheep, wandering slowly across acres of rolling grassland, were an integral part of chalk downland scenery and a traditional way of farming and rural life which stretched back over a thousand years to Saxon times.

Through centuries of constant nibbling, sheep have created a very special type of grassland, rich in unusual plants and insects which are adapted to cope with the thin chalk soils and the harsh, dry exposed conditions. In spring and summer, the remaining turf is covered with wild flowers. Harebells and orchids are scattered amongst sweet-smelling herbs such as wild basil, marjoram and thyme. Small, spreading, slow-growing plants have proved to be the most successful and there can be more than 40 different species growing within just one

View towards the eastern Downs from Ditchling Beacon

Mark Brookes

The chalk escarpment west of Devil's Dyke

square metre of undisturbed chalk grassland. Beautiful rare butterflies, such as the vivid adonis blue, thrive in these conditions. Less obvious, but equally important, are the unusual specially adapted grasshoppers, bees, ants, crickets, snails and other small animals that depend on this habitat.

There are subtle but significant differences between the plants found on the shady north-facing slopes of the escarpment and those on the warm dip-slope. The former are less rich in species but the cooler, damper conditions encourage some unusual mosses, liverworts and lichens growing in a relatively tall, coarse sward. The chalk grasslands on the southern slopes have been seriously fragmented by intensive cultivation, but it is here that the most diverse mix of species is found. The National Nature Reserve at Castle Hill has some of the richest chalk grassland in Britain.

Since the war years, when so much of the downland was ripped up by the plough, the chalk uplands have been a sharp mosaic of arable fields, with the traditional chalk grassland pastures confined to the steepest slopes - on the escarpment and the upper coombes. The habitat has become so reduced in extent that it has become an international rarity, and covers less than 5 per cent of the South Downs. Today, under the South Downs Environmentally Sensitive Areas Scheme, farmers are voluntarily withdrawing ploughing and fertilisers in the interests of conservation and the restoration of this valuable grassland resource.

Fertilisers, insecticides and ploughing are not the only threats. Neglect through lack of grazing allows coarse grasses, such as erect brome, to thrive and shade out the low, fragile chalkland species. In time, areas left ungrazed undergo a sequence of changes known as succession. First the grassland becomes tall and coarse; then bushes of hawthorn, blackthorn and dogwood become established; later still, small ash trees begin to appear and eventually the scrub layer develops into secondary woodland. It is worth remembering that the Neolithic farmers cleared just such a light woodland back in prehistoric times.

The patchy, irregular mosaic of grassland scrub on the slopes of steep coombes and the scarp is a familiar sight, lending an attractive, wild, unstructured quality to the downland and emphasising the scale and indented form of its steeper slopes. Conservationists, to protect this rare habitat, wage a constant war against the scrub, cutting it back at every opportunity. Chalk grassland is clearly the most important habitat and the invasive scrub must be kept in check, but the value of chalk scrub for wildlife is increasingly recognised and it is sometimes thinned or coppiced rather than removed so as to retain a habitat for a different range of species. Slow-growing types of scrub, such as juniper and gorse, are selectively encouraged. Bird life is particularly important. Yellowhammers, linnets, whitethroats, blackcaps and stonechats are all regular visitors; redwings and fieldfares come to feed on the winter berries.

Heathland

Heathland was initially formed when woodland cover was cleared from greensand areas and prevented from regrowth by grazing animals. The acid soils would have become progressively more impoverished as rainfall washed nutrients away and heather, the most important component of heathland, would probably have been dominant by about 500 BC.

Until the nineteenth century most of the heathland was common land, used by the poor for gathering fuel and rough grazing, both crucially important as a means of supplementing a meagre income. Heather turfs were dried as a general purpose fuel and gorse was tied into rough bundles to provide more intense heat for ovens. Such cutting, burning and grazing activities were on a small scale, but they were sufficient to remove nutrients from the heathland soils constantly, ensuring that they remained impoverished and that heather would therefore predominate.

During the past 150 years, there have been important changes in patterns of land use and management which have ended the dominance of open heathland in Sussex. Improved agricultural fertilisers have made it possible to farm marginal heathlands, reducing the remaining areas to isolated patches, often areas of common land. Areas not improved for farming were planted up with conifers to create commercial plantations. A general increase in the standard of living, and in particular the availability of relatively cheap fuels, meant that the heathlands were no longer needed to supplement local incomes.

Plant succession and a build up of nutrients in the soil led to the progressive loss of open areas of heathland. The few remaining patches only survive because they are intensively managed to prevent the encroachment of oak-birch woodland and bracken.

These lowland acid heaths are internationally important for nature conservation. The heathlands of north and west Britain have developed in relatively wet conditions and display a range of Sphagnum bog mosses, while those of continental Europe tend to be much drier and more dominated by the various heather species. The Sussex heaths fall into a significant intermediate category of humid heathland, with an unusual

Julian Gray

Iping Common

range of species. Bilberry, cross-leaved heath and yellow tormentil compete with bell heather and ling in patches of humid heath and there is a range of varying local conditions, depending on the lie of the land. Small pockets of wet-heath and valley mire are found in areas where the water table is particularly high and there are more mosses, including rare species of sphagnum.

The heathlands are teeming with insect life and support many rare species of bees, wasps, ants, spiders and butterflies. All prefer the relatively warm, sandy conditions and many show adaptations to specific types of habitat. The heath potter wasp fashions a nest out of clay pellets to form a flask which it fastens to gorse bushes with its saliva; the silver-studded blue butterfly retreats to the safety of dry ant nests while it completes its pupal phase.

The combination of warm, dry sandy ground and an abundance of insects attracts a range of heathland birds, including nightjars, stonechats, tree pipits, linnets and hobbies. The Dartford warbler occurs only on heathland; others are also found in other types of habitat but the heathlands provide an important breeding ground. Reptiles are also represented. The common lizard and the adder are abundant in drier areas, while the grass snake prefers more humid conditions. Threatened species such as the sand lizard and smooth snake are also found on some heaths.

Heathland is one of the most important, and threatened, landscapes within the AONB, important not only for its history and wildlife, but also for informal recreation and for its wild, unstructured scenery which provides such a welcome contrast to the typical neat patchwork of farmland elsewhere.

Wetland

The grazing marshes of the AONB are nationally important for nature conservation and provide a dramatic contrast to the sweeping chalk downland and wooded farmland on their margins. They are found on the floodplains of all the principal chalk valleys but are most extensive within the valleys of the Arun and the Ouse, where the floodplain broadens out to include the brooks pastures landscapes of Amberley Wild Brooks and the Lewes Brooks.

Over the centuries, these floodplains have been subjected to a constant pattern of alternate flooding and drainage as farmers and engineers have battled to turn the changing natural water levels to their advantage. In Domesday times, these river valleys were still tidal estuaries, following the slow progressive rise in the sea-level at the end of the glacial period. These marshy estuaries were gradually reclaimed through a combination of natural silting and artificial drainage. The more extensive Pevensey Levels, to the east of the AONB, were reclaimed as early as the twelfth century, but the smaller floodplains of the Sussex rivers were not tackled in earnest until later, from the fifteenth century.

By this time, rising population had forced a general expansion of cultivation, and in particular sheep farming, on the high downland. The marshy floodplains were converted to meadows to provide rich grass and clover for ewes and lambs early in spring, before they were taken up on to the downland. Beef cattle were also fattened on the pastures in some areas. The regular network of drainage channels, constructed across the floodplain to drain the winter floodwaters, is still intact. The floodwaters deposited a fine layer of silt, each year replenishing the nutrient content of the rich alluvial soils.

Attempts to drain and manipulate the river floodplains have continued ever since, with the result that they have become progressively drier. However, their importance for nature conservation has been recognised and provides some protection from declining water tables. Almost every year Amberley Wild Brooks is transformed into a spectacular sheet of water and comes alive with birds stopping to feed before continuing with their cross-country migration.

The plant life of the grazing marshes ranges from the rushes, sedges and herbs growing in the silty soils of partially drained grassland, to species more tolerant of waterlogging in wetter areas. Plants of the marshy grasslands include marsh marigold, marsh valerian and lesser pond sedge, grading to soft rush, tufted hair grass and silverweed, which tolerates periodic flooding, wherever there is standing water. The drainage ditches themselves are exceptionally rich in emergent aquatic species and are ideal habitats for many plants, such as sharp-leaved pondweed and bladderwort, which are no longer able to survive in lowland waters.

In total, 56% of all the aquatic plants known in Britain and 42% of all Britain's dragonflies are found at Amberley. This wetland is protected as a Site of Special Scientific Interest and part of it is a reserve, run by the Sussex Wildlife Trust. The range of bird life is just as impressive. Waders such as snipe, redshank, yellow wagtail and lapwing breed on the damp, tussocky grassland of the grazing marshes. In winter, migrating visitors include widgeon, Bewick swan, white-fronted geese, curlew and godwit.

There has been a gradual reduction in species, largely as a result of drainage improvements and the slow drying out of the pastures. However, just to the north of the AONB, at

John Tyler

Pulborough Brooks, the Royal Society for the Protection of Birds (RSPB) has established a nature reserve and its experiments with levels of controlled flooding have proved to be astonishingly successful: 10 threatened species returned to the area in record numbers within a very short time.

Ancient Woodland

The Sussex Downs AONB is exceptionally well-wooded. Only the open east chalk uplands have remained clear of woodland and stand out in stark contrast against their wooded western neighbour. In the Weald, it is simply a question of

John Tyler

differentiating between the type and form of woodland characteristic in any one area. Panoramic views from the crest of the west chalk escarpment reveal a sea of interwoven shades of green. The darkest are the sandstone forests of the north wooded ridges. Nearer to the foreground, the heathlands appear as extensive, irregular tracts of woodland, merging imperceptibly with the more broken patchwork of fields and woodlands of the clay vales.

Many woodlands are thought to be 'ancient', in that their sites have remained continuously wooded for more than 400 years. The oldest and most precious may be relics of the original wildwood. Others, termed 'ancient semi-natural woodland', contain only native species which have established by natural regeneration.

Today, the extensive woodlands of the western part of the AONB are an evocative reminder of the importance of trees and their management in the history of the area. Their twisted, gnarled forms and natural profusion suggest links with ancient history and they give the landscape a special intimate, secret character.

The Weald takes its name from its ancient woodland. 'Andredsweald', as the Saxons called it, remained an untamed, inaccessible wilderness long after the lighter soils of the Downs had been cleared. The sandstone ridges and deep ghylls of the north-west AONB support woodland communities adapted to relatively damp, humid conditions. For instance, the combination of sessile oak and an understorey of bilberry, usually associated with upland areas, is unique in south-east England. The more calcareous soils support oak-ash woodlands, with an understorey of maple and hazel. Much of the area has been planted with conifers or sweet chestnut coppice and the ancient woodlands form part of a richly varied, interlocking mosaic of woodland and glades. On the heathlands further south, ancient woodlands are far fewer and more fragmented. Here, conifer plantations cover substantial areas and a light secondary woodland of oak and birch, over an understorey of bracken, has grown up over the heath. An alder-willow carr, with sphagnum mosses beneath in the more acidic areas, has developed on wetter streamside soils.

The west chalk uplands to the north of Chichester are renowned for their beautiful mature woodlands. Local large estates, such as Goodwood and West Dean, take much of the credit as they have prolonged a tradition of careful forestry and woodland management. Many of the woodlands are found on the clays overlying the chalk and have an oak-hazel character, with varying amounts of birch, hornbeam, holly, ash and chestnut. They are often carpeted with bluebells and wood anemones in spring. Where the clay layer is absent, the deep chalk soils support woodlands dominated by beech, and sometimes yew. Many unusual mosses and lichens are found growing on the older trees, and in West Dean Woods a large colony of wild daffodils covers part of the woodland.

The wooded slopes of the chalk scarp have fragments of ancient woodland, but most are a combination of beech, planted by wealthy eighteenth century landowners, and naturally regenerating mixtures of oak, maple, whitebeam and yew. Towards the foot of the chalk scarp, throughout its entire length, there are traces of a particularly interesting type of ancient woodland, notable for the presence of the rare large-leaved lime. It often occurs as large coppice stools, possibly up to a thousand years old, in a mixture of wych elm, maple, hazel and whitebeam - perhaps an indication of the composition of the original woodland which covered much of the South Downs.

The woodlands are majestic all year round, but in autumn they transform the landscape into a jewelled mosaic of colour and light. The acidic soils of the heathlands bring out the true intensity of the seasonal foliage colours, but everywhere beech, chestnut and oak trees turn to rich shades of gold and bronze and stands of birch become torches of brilliant yellow. Only evergreen conifer plantations remain unchanged, providing a solid, dark green backdrop to the show.

Ancient and semi-natural woodlands are valued for their rich and varied flora and fauna. Ancient woodlands have a wonderfully diverse ground flora, with many plants specially adapted to shady conditions. Paddock Wood, for instance, has 37 examples of such rare plants (ancient woodland indicators), together with some rare lichens and mosses growing on older, decaying wood.

Almost all the woodlands in the AONB were traditionally managed by coppicing or pollarding, both techniques for ensuring a continuous supply of young wood, which was essential for the iron furnaces and charcoal camps of the Weald, and important for woodland products everywhere. In a coppice, the young stems, or 'poles', are cut every few years, leaving a dense 'stool' from which a fresh crop of poles will grow. If this pattern is repeated on a regular basis, the tree will survive for hundreds of years. Most species of tree can be coppiced, but hazel and chestnut are most commonly managed in this way. The practice allows a flood of sunlight to reach the woodland floor, encouraging a profusion of wildflowers.

Coppice woodland, which often includes larger oaks or chestnuts, is renowned for its display of spring bluebells.

Pollarding is another traditional method for preserving a supply of wood, but in this case the young stems are cut back from the top of the main trunk. This form of management was generally associated with woodlands used for wood-pasture, common land or deer parks maintained for sport by the landed gentry. The young shoots are well above the animals' reach and in this way ancient trees were preserved for centuries in woodland with a fairly open character. The ancient oak pollards in the grounds of Parham House are relics of a medieval deer park. They support a remarkable community of mosses, which includes a total of 187 recorded species, and a rich invertebrate community.

John Tyler

Stedham Common

Commons and Wastes

The commons and 'wastes' were not only a vital source of fuel and rough grazing for the poor; they also acted as a refuge for those displaced from villages or unable to buy, rent or work for land in the existing settlements. Periods of population growth and pressure on the land were traditionally marked by encroachment on the commons and heaths. However, these areas retain an unstructured, marginal appearance, a welcome contrast to the surrounding neat farmland patchwork and an important characteristic of the local landscape.

The heathland fringes have been progressively colonised for centuries by squatters' cottages and smallholdings. This type of settlement, usually built in a haphazard way, is still represented on the heathlands to the south of Midhurst and there are many isolated groups of cottages scattered throughout the woodlands of the sandstone ridges of the north-west Weald. The remaining areas of unsettled, unenclosed forest and heath represent a combination of the most inhospitable territory, common lands which have never been enclosed and former marginal fields, since recolonised by woodland. It is still possible to trace the low earthen banks, marking the boundaries of such fields, beneath the trees. Heather and bracken were burned and lime, carried from the chalk-pits of the chalk scarp, was used to improve the acid soils.

This pattern of progressive encroachment and alteration occurred throughout the AONB. An example is provided by the landscape of the Low Weald, which still carries the imprint of the woodland clearance techniques of the pioneering medieval farmers who hacked their small fields from the wildwood, leaving wide shaws between them. These have been progressively trimmed back and the field enlarged but those on the flanks of the steep stream valleys, where further encroachment has proved impracticable, are particularly well-preserved and many fields still partially resemble woodland clearings.

Commons were usually in the remoter parts of parishes, but occasionally such land was a central focus for a settlement. Villages in the Low Weald are often clustered around large greens, which originated with the colonisation of the waste by an organised group of farmers and cottagers who valued an enclosed central clearing for grazing their animals.

Historically, common land was a vital part of the rural economy and the seventeenth century enclosure of the common fields brought bitter battles for the retention and extension of land reserved for common grazing. Much common land was retained in areas dominated by a forest-based economy, but on the richer soils, where farming was profitable, land was progressively enclosed. These later enclosures of planned, neatly squared-off fields, contrast with the more irregular pattern of the earlier medieval enclosures.

Designed Landscapes

A number of England's famous grand houses and parklands are found within the AONB, including Glynde Place, Goodwood, Petworth, Firle Place and the ruins of Cowdray. There are many other smaller gems, carefully sited in sheltered, favoured spots throughout the landscape, especially in the rich farmlands of the scarp footslopes immediately to the north of the chalk escarpment.

Their estates have ensured the continuity of many traditional management practices and a diverse range of land uses, while their designed landscapes provide a rich array of special features. Spacious, sweeping romantic parklands, studded with elegant clumps of specimen trees, contrast with the surrounding farmland patchwork. Follies (such as the tower on the skyline at Uppark), avenues and designed vistas are surprise elements which have become local cultural landmarks - part of the history and folklore of the landscape.

Many estates have distinctive boundary walls, gates and lodges; entire villages of estate cottages, all with a strong unified character, contribute an ornamental quality to the landscape. Such features provide an important visible record of the history of local landownership.

Most parklands display the eighteenth century picturesque landscape style inspired by Lancelot 'Capability' Brown. Brown himself was responsible for transforming the landscape of Petworth Park in 1752 and his work there undoubtedly inspired the other local gentry to vie with each other in redesigning their parklands. Brown's influence at Petworth is evident today in the brilliant use of water in the park's waterfalls and lakes, the enormous variety of shrubs and trees and the careful modelling of the ground to create surprises and carefully designed sequences of views; a studied, more artificial effect than some of the later picturesque landscapes, but strikingly effective.

Aerial view of the planned landscape of Stansted Park

Humphrey Repton also worked in the area. He was famed for his technique of using before-and-after pictures in his 'Red Books' to explain his designs to prospective clients. He followed Brown and his work often achieved a more sophisticated style. Repton adopted the principle that "the horizon should all be wooded" and his landscapes, together with many others he inspired, must have helped to ensure the continuity of woodland cover in many parts of the AONB. It is no surprise that the woodlands of the west chalk escarpment were planted at this time, perfectly fulfilling Repton's principle.

John Tyler

Arundel Castle from the Arun valley

Repton designed the parklands of Uppark, transforming the earlier formal terraces and parterres which are still visible beneath the turf. Here, in a corner of the western downland, the designed landscapes of Goodwood, West Dean, Stansted and Uppark almost abut one another, creating a remarkable group of splendid, carefully planned landscapes which attract thousands of visitors each year.

Traditional Buildings

The vernacular architecture of the AONB closely reflects the varied geological character of the local area. A range of different sandstones have been used in the Midhurst-Petworth area, timber-framed cottages are scattered amongst the spring-line cottages of the scarp footslopes and the Low Weald has many tile-hung buildings. The more remote chalk villages and barns are uniformly constructed of flint with brick dressings. Most settlements have a lively mix of materials and the buildings of the

John Tyler

Traditional wealden building materials

AONB provide a welcoming, attractive focus in the landscape as well as a fascinating historical record.

The wealden sandstones are renowned as building materials. Those found within the AONB are a buff sandstone, from the Hythe Beds, and the coarse, grey Bargate stone; buildings using both are found throughout the western Weald and there are several intermediate forms of stone of varied textures and hues. The widespread availability of such excellent building stone encouraged a high standard of architecture and many buildings display fine examples of detailing. The estate walls and villages have a particular harmony and elegance and the local sandstones are shown to their best advantage in the vicinity of the larger estates such as Petworth and Cowdray.

The dense wealden clays were well suited to brick-making and brick buildings are found throughout the wealden landscapes, especially the Low Weald. Red brick is often combined with flint to produce an attractive, simple edging detail in the chalk downland villages.

Wood was always plentiful in the Weald and the majority of traditional cottages were built with timber frames, infilled with wattle and daub, brick or a hard, whitish local form of chalk known as 'clunch'. Over the years, the infilled sections of such walls have been repeatedly patched up or rebuilt and many of the more elaborate and weather-proof brick, half-boarded or tile-hung buildings conceal an original timber frame.

The flint buildings and walls typical of the downland settlements have a rough-hewn, solid appearance, blending perfectly with the muted tones of their surroundings. Isolated large flint barns standing in remote exposed sites are a feature of the downlands, particularly in the open east chalk uplands, where they may be the only upstanding element in the landscape for miles around. Most were built in Georgian times to house the teams of oxen which then worked the land - the manure could then be spread on the remoter parts of the Downs. The barns were also used to store some of the harvest.

The diverse range of habitats and cultural features within the AONB represent an invaluable part of the landscape heritage, contributing to its scenic beauty, historical interest and importance for wildlife and nature conservation. Together they provide a sense of continuity with the past and give the landscape its own distinctive identity.

This is an evocative, meaningful landscape, with many layers of cultural interest. It has the capacity to inspire any number of interpretations, with each individual responding in a different way. The process of sharing such a range of perceptions adds layers of richness and meaning, constantly developing and reinforcing the landscape's communal sense of value and identity. The next chapter describes the reactions of some of the AONB's more renowned residents and visitors, whose perceptive interpretations of the landscape have left a permanent and illuminating impression.

John Tyler

A tile hung estate cottage

Perceptions of the Sussex Downs Landscape

From 'Downland wildlife, a naturalist's year in the North and South Downs'. By kind permission of Mr. John Davis.

Chalk grassland at Fulking by John Davis

Over the years, the Sussex Downs have been the home and inspiration for a whole host of writers and artists. Their vision of the landscape has expanded and guided our collective perception of its qualities, helping us to see and understand it in new ways. For many, cultural associations with nationally famous literary figures such as Kipling and Belloc are of immense significance and some may have been inspired to follow in their footsteps. Artists, writers, historians, geologists and archæologists have been attracted to the Sussex Downs and their work has drawn attention to the special qualities of the AONB landscape.

Descriptive Writings

Historical accounts bring the dynamics of landscape change into sharp focus. Today it is almost impossible to imagine the tranquil forests and ghylls of the north-west Weald as an industrial heartland, the iron foundry of England. William Camden, a sixteenth century historian, provides us with a vivid description of a Weald which was, as he wrote:

> *... full of iron mines, all over it; for the casting of which there are furnaces up and down the country, and abundance of wood is yearly spent; many streams are drawn into one channel and a great deal of meadow ground is turned into ponds and mills for the driving of mills by the flashes, which, beating with hammers upon the iron, fill the neighbourhood about it, night and day with continual noise.*

Local place names such as Hammerpond House, Furnace Lane and Cinder Hill provide further piecemeal evidence that such noise and activity really happened in this seemingly forgotten corner of Sussex.

The accounts of seasoned travellers such as Daniel Defoe provide a valuable snapshot, indicating not only what the landscape looked like, but also how it compared with other parts of the country. For Defoe, writing in the eighteenth century, the Downs were clearly something special. He declared them to be "the pleasantest and most delightful of their kind in the nation".

Again and again, writers are moved to eloquence by the evocative qualities of the bare, eastern downland, inspired by their open, rolling landform. In 1772, the famous naturalist Gilbert White wrote to Thomas Pennant describing the South Downs in anthropological terms as having "broad backs" and a "shapely figured aspect". He states his preference for the smooth, open Downs rather than the more rugged landscapes of northern England. Mrs Radcliffe, a popular novelist of the day, was of like mind. Gazing across the Weald to the distant Downs, she admired how they "heaved up their high, blue lines as ramparts worthy of the sublimity of the ocean".

Such lyrical descriptions are given a firm, informed context by William Cobbett, an agriculturalist and political commentator famed for his travel account, *Rural Rides* (1853). He was impressed by the productivity and sound management of the farmland and the relative prosperity of the labourers. In 1823 he was in Singleton and he describes the route between Midhurst and Chichester:

> *[The road] goes through some of the finest farms in the world. It is impossible for corn land and for agriculture to be finer than these ... The corn is all fine; all good; fine*

The Sussex Downs - *postcard*

> *crops, and no appearance of blight. The barley extremely fine. The corn not forwarder than the Weald. No beans here; few oats comparatively; chiefly wheat and barley; but great quantities of swedish turnips, and those very forward ... There is, besides, no misery to be seen here. I have seen no wretchedness in Sussex; nothing to be at all compared to that which I have seen in other parts; and as to these villages in the South Downs, they are beautiful to behold.*

The Sussex Downs inspired some of Cobbett's most vivid descriptive writing in which he combined a real passion for the scenery with an acute understanding of the cost-effectiveness and efficiency of agricultural production.

The vast flocks of sheep which traditionally roamed the Downs seem to have inspired many writers, such as Barclay Wills in the 1930s and Tickner Edwardes, who delighted in describing the lifestyles of the downland shepherds and the daily round of rural life. However, by then downland sheep-farming was long past its nineteenth century heyday and we must return to an earlier account for an agriculturalist's view. The Reverend Arthur Young, writing in 1813, noted the importance of sheep in the local economy:

> *Between East Bourne and Steyning, which is thirty-three miles, the Downs are about six miles wide, and in this tract there are about 200,000 ewes kept: the whole tract of the Downs in their full extent, is stocked with sheep, and the amazing number they keep, is one of the most singular circumstances in the husbandry of England.*

Referring to the value of sheep in fertilising the arable land on which they were folded on winter nights, he described the Sussex sheep flock as "that great moving dung hill".

Perceptions of the landscape are based to some extent on personal preference, but are easily swayed by contemporary fashion. The nineteenth century taste for the picturesque shifted the emphasis firmly to the more romantic, wooded landscapes of the western downlands; the open chalk uplands were out of fashion and seemed dull and rather monotonous in comparison. William Gilpin, a leading exponent of the new aesthetic, described smooth, rounded landforms as "ugly". Not surprisingly, chalk downland was not to his taste. He thought that the town of Lewes was well sited, but that the views to it were spoilt by chalk "which disfigures any landscape". Similarly, Samuel Johnson described the landscape around Brighton as being so desolate that if a man had a mind to hang himself in desperation he would be hard put to find a tree on which to fix the rope.

The spa waters of resorts like Brighton and Worthing were a great attraction, but visitors were unimpressed by their downland setting. Written in the 1860s, Mrs Merrifield's *Natural History of Brighton* extols the virtues of the downland for healthy exercise and the study of natural history, but finds little to recommend in its scenery, which she describes as barren and treeless.

The rugged landscapes of the west Weald, with their precipitous slopes, deep ghylls and fast streams would, perhaps, have had more appeal, but this part of the AONB landscape does not seem to have been discovered; people went to landscapes they heard about so the Lakes and the Peak District were more favoured.

However, from late Victorian times onwards, the Downs regained their special place in the national psyche. Written at the turn of the century, W H Hudson's *Nature in Downland* (1899) was more than an inspired natural history of the Downs; it was a passionate evocation of their magical qualities. For Hudson, the Downs were a wonderful untamed wilderness - and to an extent this was actually the case, as this was a time of acute agricultural depression. Sheep-farming was in decline and the pastures were overgrown. It was from this point onwards that so many writers and artists fell under the spell of the Downs.

By kind permission of West Sussex County Council Library Service (copyright reserved)

The South Downs - *colour postcard*

Literary Connections

The poet, artist and mystic, William Blake (1757-1827) lived at Felpham, near Bognor Regis, for three years. The famous lines about "England's green and pleasant land", 'Milton: Preface' (1804-1810), are said to have been inspired by the views across the west Downs from the Earl of March pub at Lavant, just north of Chichester. For Blake, this was the quintessential English landscape: a bold sweep of field and woodland with an unthreatening, calming presence. Writers have the privilege of living wherever they choose and it is no surprise that so many have opted for this beautiful stretch of countryside.

The Poet Laureate, Alfred Lord Tennyson, lived at 'Aldworth', a house built for him on a 36-acre site high on the slopes of Black Down in the north-west corner of the Weald. His poem 'View eastward over the Weald' (1880) describes the scene from Black Down:

Over birches yellowing, and from each
The light leaf falling fast
While squirrels from our fairy beech
Were bearing off the mast;
You came, and looked and loved the view,
Long-known and loved by me,
Green Sussex fading into blue
With one gray glimpse of sea.

Tennyson clearly revelled in the spacious vistas and scenic beauty of the landscape. Cobbett, predictably, took a more pragmatic view of a landscape lacking in agricultural distinction, declaring " I have never seen the earth flung about in such a wild way as round Hindhead and Black Down."

For some, like Tennyson - and later Kipling and Belloc - the landscape itself represented a source of inspiration; for others it provided a peaceful retreat and refuge in which to work productively.

The novelist Anthony Trollope, a contemporary of Tennyson's, came to the Sussex Downs with this in mind. He leased a house in the village of South Harting, at the foot of the west chalk escarpment, where he wrote prolifically for 18 months (1880-1882), producing an impressive ream of works including *Kept in the Dark* (1882), *The Fixed Period* (1882), *Mr Scarbrough's Family, An Old Man's Love* and *The Land Leaguers.*

One who visited, and then came to stay, was the novelist and playwright, John Galsworthy. He lived and worked in the village of Bury in the Arun valley, where he wrote *The Forsyte Saga* (1922). He was awarded the Nobel Prize for Literature in 1983.

The bare, flowing forms of the eastern Downs are uniquely 'Kipling Country'. Peter Brandon (1994) explains why:

> *[Kipling] was stirred to write his verses entitled 'Sussex' and numerous others by the magnificent towering white cliffs, the sea views from the cliff tops which are such a special scenic heritage of the Downs and also the solitary places below the cliffs where one can see and hear the surf crashing against the rocks. Kipling skilfully captures the essence of the intangible atmosphere of the Eastern Downs that has since been lost, such as the voice of the shepherd, the barking of his dog, the cries of the sheep, the far-off clamour of the sheep-bells, jingling of harness, calls of the birds and the sound of the sea, in the absence of any mechanical noise whatsoever, something not to be heard any more and yet something so simple and familiar to downsmen since the very beginning of man's farming the Downs some six thousand years ago.*

To quote from Kipling's poem, 'Sussex' (1902):

No tender-hearted garden crowns,
No bosomed woods adorn
Our blunt, bow-headed, whale-backed Downs
But gnarled and writhen thorn -
Bare slopes where chasing shadows skim
And, through the gaps revealed,
Belt upon belt, the wooded, dim,
Blue goodness of the Weald.

While Kipling's evocative verses made the open East Sussex Downs famous throughout the world, Hilaire Belloc, also writing at the turn of the century, put the wooded Downs and Weald of West Sussex firmly on the literary map.

Born in France, Belloc grew up in the village of Slindon and adopted Sussex as his spiritual home. He was a poet and a

Burpham Village - *colour postcard*

John Tyler

Swanbourne Lake

romantic, inspired by the scenery and tranquillity of the rolling field and forests. In 'The South Country' (1910), he wrote:

> *I never get between the pines*
> *But I smell the Sussex air;*
> *Nor I never come to a belt of sand*
> *But my home is there.*
> *And along the sky the line of the Downs*
> *So noble and so bare ...*

Belloc also studied the landscape in a scholarly way, enthusiastically seeking to understand its geology, landforms and history. He has been much maligned for some drastic historic inaccuracies, but his *The County of Sussex* (1936) is nevertheless a charming guidebook, full of local anecdotes and details which make the landscape come alive with interest. His descriptions of "bunches of pine trees, making a peculiar note in the landscape" along the heaths to the north of the scarp, and of "the hollow green roads ... long waterlogged" of the Weald and the "wooded mass of the Nore Hill, now uninhabited and silent, but once a stronghold", capture the details of familiar views and moods. The distinctive characteristics of the landscape cannot have changed all that much since the 1930s, as the places described in Belloc's prose are instantly recognisable.

The railways, and soon after the motor car, made the Downs more accessible to visitors and the growing popularity of natural history, geology, archæology and botany encouraged a wave of enthusiasm for rambling and rural exploration. Writers such as Richard Jeffries, and later W H Hudson, introduced scores of people to the natural history, landscape and rural culture of the Downs. Peter Brandon (1994) points out:

> *Jeffries is distinguished from most nature writers by his interest and knowledge of the farming which underlay the downland scene where his preference for nature was interwoven with man's agricultural activities. His sense of the past was quickened by the prehistoric tumuli and the ancient field systems then imprinted over the higher slopes of the South Downs and he is the earliest writer to perceive the history of this visible ancient land as a continual process of evolution in such forms as the shapes of fields, the line of trackways, the traditional crafts, the old implements such as ploughs, and wagons, and the names of villages, thus anticipating the modern interpretation of the English countryside as a palimpsest of human decisions by W.G. Hoskins.*

While the wealden landscape seems to have avoided public notice, the South Downs received a good press at the turn of the century and into the 1920s and 1930s. The popular market was flooded with guidebooks and articles. These include E.V. Lucas's *The Highways and Byways of Sussex* (1904), Arthur Beckett's *The Spirit of the Downs* (1909) and A Hadrian Allcroft's *Downland Pathways* (1924). Images of the South Downs became associated with summer holidays and escape from city life - an idealised picture-postcard landscape. Writers' and artists' colonies proliferated, notably at South Harting, Amberley, Storrington, Ditchling, Rottingdean and Rodmell.

Some became more permanent residents. Virginia and Leonard Woolf lived at Rodmell, in the Ouse valley, for over 20 years. The beauty of the downland around the village gradually permeated Virginia Woolf's writing, particularly her diaries and last novel *Between the Acts*, which is set in the nearby village of Alciston. She became increasingly in awe of nature and of her own inferiority and inability to express herself in such surroundings:

> *One is overcome by beauty more extravagantly greater than one could expect ... I cannot express this, I am overcome by it, I am mastered.*

The artist Vanessa Bell, Virginia Woolf's sister, lived close by at Charleston, where she encouraged extended visits from a rather eccentric group of writers, art critics and painters, including her husband, Clive Bell, artist Duncan Grant, writer David Charleston and critic Roger Fry. E M Forster, the Woolfs and Cyril Connelly were frequent visitors and the house became the acknowledged summer retreat of the Bloomsbury set.

Over the years, the Sussex Downs have been the setting for many novels. The best known, and still one of the most successful, is Stella Gibbons's *Cold Comfort Farm* (1932), a wicked, satirical jibe at the trite, postcard image of rural life. The blighted farmhouse, owned by the Starkadders, is on the bleak hillside, "fanged with flints", near Brighton.

Predictably, the Sussex landscape provides a more mellow, rosy romantic setting for many Victorian novelists, including R D Blackmore, who based *Alice Lorraine: A Tale of the South Downs* (1875) on a landscape "westward of that old town Steyning, and near Washington and Wiston".

Today a local farmer, Richard Masefield, continues the tradition of the downland novel. His first three novels have all been inspired by the chalk scenery surrounding his farm and their titles are taken from the summer butterflies of the Downs: *Chalkhill Blue* (1982), *Brimstone* (1987), a smuggling adventure set in 1773 on the Sussex coast, and *Painted Lady* (1988).

Music

It is perhaps not surprising to find that several composers have been drawn to this peaceful countryside to live and work. Those best known include Sir Arnold Bax (1883-1953), who lived at Storrington, and Edward Elgar, who lived for a time at Fittleworth and occasionally played the church organ there. He wrote his *Quintet and Cello Concerto* (1919) while staying there. However, relatively few landscapes have been the direct inspiration for great music. Local composer John Ireland (1879-1962), who lived at Rock Mill, Washington, stated that "much of my music has been inspired by the beauty of Sussex and its ancient past".

Ireland wrote *A Downland Suite,* but *Legend*, composed as a direct experience of the landscape near Burpham, and *Amberley Wild Brooks* are said to represent the "supreme expression of this inspiration" (*Sussex County Magazine*, vol. 29, pp. 63-66).

Today, the Downs are still a source of inspiration for modern bands living around the area.

Visual Arts

Both Turner and Constable worked in West Sussex, invited to boost the prestige of the wealthy landed gentry. Turner was a regular visitor at Petworth in the early 1800s. His many pictures of the house and its parkland include 'Petworth House from the Lake; Dewy Morning' (1810) and 'Petworth Sunset Across the Park' (*c.* 1830). Constable visited Arundel in 1834 and 1835 and 'Arundel Mill' (1837) is believed to be his last work. His fresh watercolour and pencil sketches of 'Cowdray House: The Ruins' (1834) and 'Tillington Church' (1834) were executed during day-trips to the surrounding countryside while staying with Lord Egremont at Petworth in September 1834.

Painting at the same time, and probably influenced by Turner, the artist Copley Fielding captured the essence of the downland landscape with an innovative, fresh style. His atmospheric watercolours conveyed the spacious expanses of the Downs and the dynamic play of mist and light. Another artist who excelled in his ability to convey the dignity and beauty of the Downs was Philip Wilson Steer, who painted his watercolour 'Sussex Downs' in 1914.

In 1910, the *Sussex Daily News* reported that 36 artists listed as resident in Sussex had exhibited their work at the Royal Academy Summer Exhibition, including Henry Herbert, La Thangue of Graffham, Edward Stott of Amberley and Charles Sims of Lodsworth. The trend continued throughout the 1920s and 1930s, when numerous artists came to live and work in this part of Sussex. Among them were Stanley Roy Badmin, who lived at Bignor, Sir Frank Brangwyn at Ditchling and Edwin Harris at Storrington.

By kind permission of the Charleston Trust

Charleston Pond 1916 by Vanessa Bell

Eric Ravilious's prodigious talent flourished in Sussex. Taught by Paul Nash, he experimented with new ways of depicting the chalk downlands. Many of his paintings are of a cottage he often visited called 'Furlongs' and its downland setting. He had a refreshingly unsentimental approach. 'Waterwheel' (1934) combines wire fencing and agricultural machinery with meticulously drawn downland lines and he also painted the dazzling white surfaces at the Asham Cement Works. Peter Brandon (1994) admires "his ability to depict in a three-dimensional form the massively rounded shoulders of the chalk and the great spaces of the coombes between, almost as sculpture".

Concentrating on the wooded landscapes to the north of the Downs, Ivon Hitchens (1893-1979) bought a piece of woodland at Lavington Common near Petworth in 1940, after being evacuated from his Hampstead home because of bomb damage. He had painted in West Sussex throughout the 1930s and chose to return on a permanent basis. He built a house, created a lake and later acquired an orchard. His garden became the subject for his work thereafter, but he also found inspiration in its surroundings.

His pictures 'Curved Barn' (1922), featuring Bex Mill at Heyshott, and 'Sussex River near Midhurst' (1965) are at Pallant House Gallery, Chichester.

Nestling in the shadow of the chalk escarpment, the village of Ditchling became famous for its group of artists. Led by artist, sculptor and engraver Eric Gill, the group became immersed in a back-to-the-land movement (the typeface Gill Sans is used within this document). Edward Johnstone, an innovative calligrapher and Douglas (Hilary) Pepler, printer and publisher, lived communally with Gill in a close-knit religious community. Others came to join them and collectively they made a gesture of defiance against the politics of urban industrial mass-production. They built a Dominican chapel and a group of cottages and workshops on the edge of Ditchling Common in 1913. The Guild of St Dominic that they founded there was not wound up until 1989.

The landscape of the Sussex Downs continues to attract and inspire artists such as Adrian Berg, Jeffrey Camp, Norman Clark and Douglas Gray. Bill Brandt's 'East Sussex Coast' (1953) is an evocative experiment, combining the flowing, distorted lines of a nude figure with photographic imagery of the chalk coastal cliffs to create a new imaginary landscape which powerfully recalls the gentle, feminine forms of sweeping downland. Peter Brandon (1994) considers that Carol Weight's "strange and haunting individual interpretation of the West Sussex downland scene is one of the major artistic

Downs in Winter c.1934 by Eric Ravilious

developments at the present time". Roger Coleman (1981) explains how his close relationship with the local landscape influences his work:

> *Standing on a chalk scarp above the River Arun, people seeing Burpham for the first time invariably describe it as beautiful, but compared with many Sussex villages the architecture is not outstanding ... what makes it attractive is its landscape; not simply the Downs, but the closer landscape of the situation of the buildings collectively in their setting. So Burpham emerges as a small village of unselfconscious beauty and astonishing tranquillity; one of those rare places where time can appear to pass slowly and without complication.*

Carolyn Trant's recent drawings and prints of earthworks on the Sussex Downs reveal the layers of cultural history imprinted on the chalk. Her work depicts contemporary mountain-bike tracks alongside the ramparts of hillforts, uniting time and space in a historic continuum and emphasising that this is a landscape fashioned by centuries of human activities and that its future lies in human hands.

Each of the different landscapes of the AONB has its own special appeal and its particular exponent. Kipling is unrivalled as spokesman for the open east downland and Belloc is particularly known for extolling the romance of the west wooded Downs and the Weald. Many have followed in their footsteps, providing a fascinating and valuable record of the AONB landscape and our evolving perception of it. It is perhaps significant that many contemporary artists have chosen to focus on its relative fragility and vulnerability in the face of forces for change. Agriculture, development pressures, mineral extraction, and tourism are all perceived as threats.

Returning to more concrete evidence, the next chapter assesses the impact of these forces and the extent to which their negative influence can be mitigated.

By kind permission of Carolyn Trant

Mount Caburn Hillfort with Fossil Sea Urchins 1990 by Carolyn Trant

Forces for Change

John Tyler

Hang-gliding on the chalk escarpment

This chapter examines the major forces for change which are likely to influence and shape the landscape of the Sussex Downs AONB. Clearly, some changes will apply to the whole of the AONB, while others are likely to be specific to particular parts. The sensitivity or robustness of the different landscape types will determine the impact of the 'force' and whether adverse consequences may result. Change, in itself, is not necessarily harmful, but its consequences do need to be understood. The overall aim should be to maintain or enrich the characteristic features which contribute to the landscape's distinctive identity and local sense of place.

All landscapes are constantly in a state of flux, evolving and changing as a result of both natural and man-made processes. Rates of change vary. Some, such as the gradual erosion of the South Downs themselves by wind and rain, are so slow as to be barely perceptible, whereas the process of erosion along the coast at the Seven Sisters is very much in evidence as pieces of the cliff crack and crumble away at a rate of up to one metre a year.

However, the changes man has wrought, through working the land and establishing settlements, tend to have a far more immediate impact on landscape character than natural evolution. The pace of change in the twentieth century has been far greater than ever before.

There is no reason to believe that the forces for change will slacken off in the foreseeable future. The designation of the Sussex Downs AONB recognises the particular vulnerability of this exceptionally fine landscape in the face of ongoing development and seeks to conserve its qualities. To some extent, this is a protected landscape, but only to a degree - the situation requires constant monitoring as the processes of change, if left unchecked, can lead to the degradation of landscape quality. Local identity, ecological diversity, historic remnants and a sense of remoteness are easily eroded, often in an *ad hoc* manner, as a result of apparently minor changes.

To the west, the Sussex Downs AONB abuts the East Hampshire AONB and includes not only the South Downs but also the high greensand ridges to the north. The entire area sheltered by these two dominant landforms therefore has protected status. However, to the east of the Arun valley, the AONB boundary runs close beside the foot of the chalk escarpment, restricting the protection to a smaller area. Nevertheless, the panoramic views out to and beyond the designation boundary from the crest of the chalk escarpment, together with those from outside looking towards the 'wall of the Weald', make a crucially important contribution towards the special quality of the downland landscape. Similar issues arise along the southern margins of the AONB. Here the extensive urban development along the coastal plain just outside the designated boundary has a significant visual influence on the scenic character of the open chalk hills. These important visibility issues should be borne in mind in assessing the forces for change; clearly areas beyond the boundaries of the designated AONB should be addressed, as well as those within.

Agricultural Change

From Neolithic times, when the woodlands covering the chalk downland were first cleared, to the combine harvesters of the twentieth century, the most significant force for change in this rural landscape has been agriculture. Until very recently, the emphasis had always been on increasing productivity: the eighteenth and nineteenth century 'improvements' of the Agricultural Revolution and the tremendous push to maximise food production during the war years, when acres of downland pastures were ploughed. At that time, political means such as compulsory cultivation orders and grants were used to control patterns of land use. That particular crisis is long over and the market forces operating today are very different. The control of agricultural change is still very much in the hands of the politicians, but they are now becoming motivated by the need to promote sustainable environmental management and to avoid surpluses rather than productivity.

The most important decisions are no longer taken at a local level - the national, European and wider international policy framework has more influence. This political and economic background has a strong effect on the viability of different types of farming and therefore the appearance and vitality of the countryside. The current climate is one of change and uncertainty for agriculture. However, there are potentially more opportunities for guiding changes in land use and land management that are sympathetic to the special qualities and character of the area.

Two examples are particularly pertinent. In recent years, reforms in the Common Agricultural Policy have required farmers to 'set-aside' a proportion of their arable land. This has had a particularly significant impact on the appearance of the chalk uplands, with blocks of land being abandoned. Market forces also have an influence and have resulted in a more diverse range of crops: rape and linseed now add seasonal splashes of brilliant colour to the otherwise muted tones of the chalk farmland.

On a larger scale, and of particular influence on the chalk landscapes, is the South Downs Environmentally Sensitive Areas (ESA) Scheme, which aims to encourage the traditional farming methods that have helped to create the landscape character, wildlife habitats and historic features of the chalk hills and valleys. The Scheme is voluntary, but farmers are given financial incentives to undertake specific forms of management over 10-year periods. The South Downs ESA was originally designated in 1986 and local farmers have responded well. The result has been an encouraging increase in the management of previously neglected chalk grassland slopes and, in particular, the conversion of arable land to pasture. While they can rarely match the floristic diversity of true species-rich chalk grassland in ecological terms, these recently converted pastures have had a significant visual impact. The change is particularly evident towards the escarpment, where such grassland has most ecological and visual influence because there are often opportunities to link with existing areas of unimproved chalk grassland. An added benefit is that many archæological

John Tyler

Sheep grazing on chalk escarpment at Firle

Arable crops at the foot of the Downs

sites are now protected under the ESA Scheme, though many have already been destroyed by post-war ploughing.

While much of the countryside shows signs of increased mechanisation and intensified agricultural production in the form of hedgerow loss and groups of large, steel-framed farm buildings, some of the more marginal land has suffered from neglect. Small farm woodlands are rarely coppiced, thinned or managed and a reduction in stock has allowed scrub to creep over the steep slopes of the chalk scarp, threatening the remaining unimproved chalk grassland.

The rate of hedgerow loss has slowed, but has not ceased. Lack of management and loss of hedgerow trees are the main threats. In many areas, particularly the landscapes of the western Weald, these important characteristic landscape features are not being renewed once they have declined. In many instances, die-back is accelerated by damage from road spray, agricultural machinery and road-widening schemes. Many trees are not healthy and so are susceptible to disease.

Apart from the changing influence of market demand and land use policies, increasing pressures for profitability in the face of a competitive market are encouraging significant changes in the structure of farms and the distribution of land. In some areas, such as the chalk uplands, farms are being amalgamated to allow more efficient production, while in others, particularly the west chalk valley systems and the scarp footslopes, there is evidence of subdivision.

The proliferation of smallholdings and horse pastures often leads to a poorer quality of management. Everywhere the pressures to supplement profits from farming are seen in the diversification of farms, with 'bed and breakfast' signs, small caravan sites and barn conversions. On many estates mineral working, particularly for sand deposits, has become an important additional source of income.

Sussex FWAG

Woodland management

Generally the large estates have provided an element of stability, maintaining traditional craftsmanship and a diverse local rural economy with an emphasis on both farming and forestry. The persistence of the historic pattern of land ownership has allowed small tenant farms to continue productively and has encouraged the maintenance of traditional farm buildings. The historic parkland landscapes associated with the estates have often suffered from intensive agricultural methods (for example, intrusive game-crop patches) but generally the presence of so many large estates has represented a very positive force for the conservation of local landscape character and distinctiveness in the AONB.

Woodland Management

The woodlands of the Sussex Downs AONB contribute enormously to the quality and character of its landscape, particularly in the west, where woodland covers approximately 60% of the area. The woodlands vary in structure, species composition and form and each landscape type is characterised by a particular combination of woodland, farmland and settlement. For example, in the north wooded ridges landscape, an interlocking mosaic of different types and structures of woodland is broken by scattered open glades of heath or rough grazing, whereas in the Low Weald the patchwork of pastures is separated by winding shaws and small blocks of woodland.

While the commercial forestry plantations managed by Forestry Enterprise and the estates are profitable, smaller farm woodlands, copses and shaws are generally no longer part of the rural economy and there has been a general decline in the quantity and quality of broadleaved woodlands. Apart from the lack of economic return, the lack of knowledge of woodland management and the decline in the traditional local skill-base are becoming serious constraints.

The traditional management techniques of coppicing or pollarding, thinning and selective felling of older trees created a diverse woodland mosaic, characterised by stands of trees of different ages. Small stakes and coppice poles were regularly harvested and used for a whole range of products including bean sticks, fencing, charcoal, fuel, wood pulp and household implements; larger timbers were used for construction or furniture. The decline in small woodland management is due mainly to the loss of such small-scale local markets in the face of cheaper imports and alternative materials. Only the larger timber has value, and the result, in terms of woodland management, has been insufficient thinning and therefore too many poor quality trees, all of the same age.

Such stands are particularly vulnerable to disease and windthrow. Storm damage, as a result of the 1987 and 1990 gales, has exacerbated the problem. Small woods in exposed locations are especially at risk and if they are relatively unmanaged, with an even age structure, they are likely to be permanently devastated. Special grants and incentives have not proved sufficient to replant and manage storm damaged woodlands when they are in relatively inaccessible places. Unfortunately, such woodlands are often on steep slopes and upland areas where they have an important visual impact and are recognised local landmarks.

On a more positive note, there has been something of a revival of interest in traditional forms of woodland management such as coppicing, and the estates, with their long history of careful woodland management, have preserved a rich heritage of small, healthy woodlands in some parts of the AONB. Changes in grant aid have also helped: the Woodland Grant Scheme (introduced in 1991) recognises the contribution of broadleaved woodland to the landscape and the environment generally, and the Farm Woodland Scheme provides incentives for woodland management and planting as part of the farm economy.

Following the Second World War, many relatively derelict woodlands, together with marginal heathlands, were planted up with conifer trees. They remain as part of the woodland mosaic, but rarely dominate the landscape over extensive areas. There is ongoing pressure for the conversion of broadleaved woodlands to more profitable, short-term conifer plantations, but generally there has been an improvement in the quality and appearance of commercial forestry within the AONB. The commercial plantations are becoming more diverse and more 'environmentally friendly' in character, with an increased recognition of their value both in scenic terms and as a recreational resource. For example, Friston Forest has been progressively replanted with broadleaved species. However, there is plenty of room for improvement, particularly in creating more sensitive relationships between the edges of plantations, landform and the scale and character of the surrounding landscape pattern.

Today, few woodlands are 'in balance' with nature. Exotics such as rhododendrons, laurel and Japanese knotweed are a constant problem. These species spread rapidly from nearby gardens and parkland. They thrive on the acidic, sandy soils of the greensand ridges and shade out the more diverse range of native plants. Once established, these species are almost impossible to eradicate, but ongoing management can at least keep them in check. Grey squirrels, another introduced species, are an additional menace, selectively stripping bark from mature deciduous trees. They tend to leave scrubby species such as elder and hawthorn so the development of true woodland is hampered. Fallow deer and rabbits may also 'do too well', preventing natural regeneration by eating young saplings.

John Tyler

The A27 north of Brighton at night

Development Pressures

The AONB designation confers an additional degree of protection in relation to development control. The prime objective is the conservation of the natural beauty of the landscape: large scale residential, commercial or industrial development is unlikely to be permitted. However, such conditions do not apply in areas which lie beyond the AONB boundary, but which are visible from the chalk escarpment.

The visibility of development in such areas from the relatively narrow eastern section of the AONB is particularly pertinent. Pressures for urban development immediately beyond the AONB boundary to the south of the Downs have led to a very abrupt interface between the smooth, open rolling downland of the AONB and the extensive urban conurbations on the coastal plain to the south. Associated urban pressures such as pylons, increased traffic and golf-courses are particularly intrusive in this characteristically open, expansive landscape. Most recently, the Brighton bypass has extended the urban influence further into the AONB, eroding its valuable sense of remoteness and isolation.

The pressure for development on the fringes of the AONB is likely to remain high in the foreseeable future, particularly

John Tyler

Traditional and modern farm buildings

around the coastal towns, where land available for development is becoming increasingly scarce. It is therefore important that a thorough, detailed visual analysis of this rural-urban interface is carried out before any form of development is even considered and, wherever possible, the impact of existing, visually intrusive elements should be reduced.

The landscape of the AONB itself is almost entirely rural in character - only two relatively small market towns, Petworth and Midhurst, fall within the boundary. Both of these towns, and some of the larger villages, have some limited suburban development which sometimes seems crude in relation to the typical vernacular architecture and the understated rural surroundings. The more extensive ribbon development within the deep, linear valleys immediately to the south of Haslemere on the north-western boundary of the AONB is relatively well screened by trees and landform, but the sense of mystery and secrecy which prevails in other parts of the north wooded ridges landscape is completely lost in this area.

Some forms of 'appropriate development' are permitted within the AONB, generally only when the proposals are on a small scale and essential to meet local community need. One particularly relevant example in this rural landscape is the conversion of farm buildings which have become redundant as a result of changes in agricultural practice. In the Sussex Downs AONB, such traditional farm buildings are often important local landscape features and their future is therefore a cause for concern. While it is important to encourage re-use of such redundant buildings so as to maintain their external fabric and appearance, their conversion to residential dwellings is often counter-productive as new windows, curtains, power lines, gardens, etc are inappropriate. As a result, re-use for non-residential uses is often encouraged. Such use may have the additional benefit of providing local employment opportunities. The option of leaving such buildings unused may also be worthy of consideration.

The construction of new farm buildings has been a significant force for change in the AONB, running alongside the decline in use of the older, traditional buildings as they have become too small or too inaccessible for modern farm machinery. In most cases, these buildings are not designed or sited with regard to their visual impact on the wider landscape, decisions being based on the operational requirements of the farm.

Small-scale residential developments may be considered 'appropriate' where they are necessary to meet local need. Such schemes may represent the largest building project in many of the villages of the AONB and thus it is important to ensure that their design maintains local character. It will also be important to consider the cumulative impact of a number of individual, similar buildings in the same area - such as stable blocks - when applications are received separately.

Where development does take place within the AONB, it is essential that the character of the landscape is maintained and if possible strengthened. Appropriate development can complement the landscape if the following principles are observed:

- employing the highest standards of design and ensuring that proposals are in sympathy with local architectural and landscape character;

- ensuring the sensitive siting, scale, form and massing of development; and

- careful choice of colour and materials.

A wide range of different forms of settlement and local building materials contribute to the special character of the landscape and criteria for the siting and design of new buildings will vary between the different landscape types within the AONB.

Despite stringent statutory planning policies for development control, there is a form of development pressure within the AONB which is less easy to control. Creeping suburbanisation occurs when individual properties are embellished with 'urban details'. Examples are all too familiar: neatly mown verges edged with white bollards and chains; over-scaled gate-posts; Leylandii hedges; elaborate fretwork on walls and fences, etc. Such suburban elements seem out of place when imported into a deeply rural landscape like the Sussex Downs. Even minor details have a cumulative effect and quickly erode the identity of the local landscape and, in particular, the important relationship between groups of traditional buildings and their immediate surroundings.

Infrastructure

Proposals for the construction and improvement of new roads and services within the AONB represent one of the most significant forces for change. The Department of Transport (Highways Agency) has recently put forward three trunk road schemes within the AONB. Following a lengthy Public Inquiry a decision on the A27 Worthing - Lancing Improvement is awaited. A preferred route was identified in 1993 for a new dual carriageway road between Lewes and Polegate. In November 1995 the Department announced that the scheme would be reviewed as a potential smaller scale improvement. Earlier proposals for a new A26 road between Beddingham and Itford Farm (Newhaven) had already been withdrawn and the project has now been dropped.

The scale of dual-carriageway roads and associated engineering structures mean that such roads are imposed on the landscape with a significant visual impact and other adverse environmental consequences.

The recently constructed section of the Brighton bypass illustrates many of the issues. In places the road forms a scar in the landscape, slicing through ridges and woodlands. Some of the associated bridges, culverts and junctions are particularly visually intrusive. In addition, increased traffic noise and road lighting pervades a large area of otherwise quiet countryside. The bypass corridor is likely to be the focus of development pressures and will generate a demand for motorist facilities. Since the bypass is almost entirely within the AONB, acceptance of such development will inevitably further erode the existing character and quality of the area.

On a smaller scale, minor road 'improvements' such as straightening 'dangerous' sections, the introduction of mini-roundabouts and white lines, road lighting at junctions, laybys and kerbing all add up to give a homogenising influence which threatens and distracts from local landscape character. Visitor pressure, particularly during the summer months, leads to congestion and illicit parking in narrow roads and it is easy to see why there is a temptation to upgrade. However, narrow twisting lanes, high hedgebanks and steep gradients are all important local features which are vulnerable to insensitive improvements. Alternative solutions may be required. Perhaps the use of cattle grids and sharp bends or bumps to slow traffic and reduce the attraction of using the narrow downland roads as commuter 'rat-runs' between settlements in the Weald and the coastal towns. Careful siting of car-parks, provision of information, promoting alternative forms of transport and better integration of public transport may all help. In some places, a sense of remoteness and tranquillity may only be preserved by more radical steps such as road closure.

A recent spate of new telecommunication masts accompanied the development of new national networks by the two personal communication network operators - Orange and Mercury. The new equipment is now largely in place and further changes are not likely in the immediate future, although prediction is

John Tyler

The A27 over the Adur valley

John Tyler

Electricity pylon in the downland landscape

difficult and it would be wise to monitor the changing requirements and to remove equipment as soon as it becomes redundant.

The recent Countryside Commission initiative to secure a national commitment to undergrounding overhead electricity cables (CCP 454) may make a very positive contribution to the quality of the AONB landscape. As in all parts of the country, the AONB has a substantial network of overhead lines, most of which have been in place for many years. Those on the expansive open downland landscapes are particularly conspicuous and should be a priority in any future programme for the undergrounding and re-routing of overhead lines. On a smaller scale, the removal of the clutter of low-voltage lines within villages - which are often conservation areas - would enhance the local environment for the benefit of residents and visitors alike.

Minerals

There is a long history of mineral extraction in the AONB. Chalk has been quarried from the Downs for centuries and the concentration of chalk pits along the chalk escarpment reflects the demand for chalk to produce lime for fertilising the farmland of the Weald to the north. The oldest pits have been well integrated into their surroundings and some, such as those on the valley slopes near Amberley, are recognised local landmarks. However, the largest quarries dating from the nineteenth and twentieth centuries, such as at Cocking, are less subtle in their impact and create dazzling white scars on the chalk escarpment and slopes of the principal chalk valleys. There is a concentration of such sites in the Lewes area, and the quarry and associated derelict cement works in the Adur valley can only be described as an eyesore.

In addition to chalk, the sandstones of the Folkestone Beds, within the heathland mosaic landscape type, have been extensively quarried for building sand. The sand quarries within the AONB are generally well screened by the surrounding woodlands but unfortunately the extensive quarries near Washington, which lie just beyond the AONB boundary, are highly visible in views from the chalk escarpment.

It is likely that quarrying for both sand and chalk will continue in the AONB in the future. However, the indications are that government policy is becoming stricter in relation to the environmental impact of quarries within protected landscapes and it is significant that MPG6 (published in April 1994) places the assessment of mineral development in AONBs on a par with National Parks, requiring that developments should be in the public interest before being allowed to proceed. Such policies only apply to aggregates (sand and gravel) and not chalk, but it is a step in the right direction. If additional mineral allocations are made in the AONB, it will be essential to test their potential impact on landscape character and to specify clearly at the outset how sites are to be restored.

Given the number of existing redundant quarries - and the fact that several others are likely to be worked out in the near future - there are important opportunities for creative restoration schemes which take full account of the site's potential for geological interest and nature conservation, as well as their wider visual impact.

Recreation and Tourism

Ever since the arrival of the first railway link to London, the South Downs have been a mecca for tourists, weekenders and day-trippers, all understandably seeking to escape from the city and enjoy the natural beauty of the countryside. In addition, the area provides an important recreational resource for the urban areas along the south coast and the major inland towns such as Guildford and Crawley.

The many historic houses and parks, such as Goodwood, Petworth and Firle Place, and popular sites such as Devil's Dyke and Ditchling Beacon, are important attractions which tend to create 'honey pots' of intense visitor pressure. In some cases (particularly sites which are natural viewpoints), the visitors are a threat to the very qualities they are seeking - a sense of space, tranquillity and the opportunity to 'get away from it all'. Car parking is a constant problem and dogs may be a threat to sheep and deer. The pressure of so many people inevitably leads to problems of litter, noise and the erosion of paths. Small informal car-parks, designed to blend with the character of the surrounding landscape when not in use, are one solution, although they have sometimes proved to be a temptation for petty criminals.

New forms of recreation, such as mountain-biking, hang-gliding and four-wheel-drive vehicles, may not be compatible with quieter pursuits and all tend to increase levels of noise and disturbance. Golf-courses are not new, but they are particularly prevalent on the open chalk uplands to the north of Brighton, where there are six separate courses, each with its own club house and car-park. Those on ridgetops, such as the golf course near Devil's Dyke, are especially prominent, but all have a homogenising influence, with the predictable scattering of flags, bunkers, lush greens and signposts. There is considerable scope to encourage environmentally sensitive golf-course management, with areas of regenerating scrub, rough grassland and fairways with a less irrigated, artificial appearance.

John Tyer

Walking in the Downs

Policies for the conservation and enhancement of the AONB landscape should recognise the fact that it is an evolving resource. It is essential that any potentially damaging forces for change are carefully monitored and tempered to mitigate their impact. National policies, such as the ESA Scheme and grants for woodland or heathland management, are often voluntary but they have an important role to play. The attitudes of farmers, land agents, landscape managers and developers are clearly crucial but we are all responsible for the landscape heritage and its conservation will ultimately depend on the influence of all those who live and work in the AONB as well as the many who visit.

The final chapter traces the inspiring efforts and successes of those who recognised the value of this landscape in the past and campaigned for its protection. It summarises the reasons for the designation of the Sussex Downs AONB as a nationally important landscape and outlines a practical strategic vision for its future conservation.

John Tyler

Parascending from the chalk escarpment

The Importance of the AONB Landscape

Peacehaven - 1930s aerial view

The Battle for the Downs

The South Downs were the cradle of the town and country planning movement in England. Most of the key issues relating to development control were first played out in the battle to conserve the stretch of downland around Brighton and Eastbourne in the face of creeping urban sprawl and the lack of legislation by which to control it in the 1920s.

In 1923 the Society of Sussex Downsmen was founded in response to serious public concern for the conservation of the Downs. It aimed to coordinate the various associations and societies with an interest in preserving the rural amenities of the Downs and to educate the public towards a better understanding of this important landscape. The Society of Sussex Downsmen and to a large extent the national Council for the Protection of Rural England (CPRE) were spurred into being by the development of Peacehaven, a geometric grid of makeshift streets and cheap housing plots offered for speculative development on the cliff tops to the east of Rottingdean. It was essentially a shanty town, initially without made-up streets or even sewerage. The Society of Sussex Downsmen, the National Trust, the CPRE and East Sussex County Council collaborated in their efforts to purchase, secure and conserve the eastern Downs for the twin purposes of farming and quiet recreation. However, an extensive tract of farmland, stretching from Brighton to the Devil's Dyke, was owned by the town of Brighton and, while it was zoned to remain 'unbuilt', a spate of speculative proposals for a 'super' racing track, hotels, golf-courses, roads, etc made it clear that the area was to be made available for a whole range of leisure facilities. The dispute which followed was to prove eventful in the history of the Sussex Downs, and of town and country planning in Britain.

The immediate outcome was the proposed Sussex Downs Preservation Bill, which broadly recommended the strategic preservation of the Downs. However, it included some particularly stringent restrictions which proved too rigid for Brighton and some of the local landowners. The Bill was rejected but the die had been cast and the struggle for the preservation of the Downs entered a new phase.

Mark Brookes

Walkers resting on the annual South Downs Way walk

East Sussex County Council aimed to preserve the entire downland above the 300 ft (90 m) contour level and conducted negotiations with individual landowners to secure agreements which restricted the uses to which land could be put. These negotiations were remarkably successful and more than 85% of the downland covered by the abortive Bill became protected by such agreements (Section 34 Agreements, under the 1932 Town and County Planning Act). Numerous potential developments of new settlements, resorts and roads were thwarted in this way during the inter-war years, before the stricter statutory land use controls of the Town and Country Planning Act were introduced in 1947.

By the 1940s the conservation organisations had turned their attention to the changes wrought by the intensification of agriculture. The ploughing of the downland was seen as a tragedy for the history and character of the local landscape but, more importantly, it threatened the recommendation, by John Dower and Sir Arthur Hobhouse, that the South Downs be considered suitable for National Park status. The areas meeting the criteria for designation were defined as being "an extensive area of beautiful and wild country" which afforded opportunities for public open-air enjoyment.

While the Sussex Downs must have felt sufficiently 'wild' to merit National Park candidacy when it was visited by the Committee in 1947, the ongoing ploughing in the years that followed dispelled this vision. In 1957 the National Parks Commission, the body responsible for designating National Parks, stated that since the Hobhouse Committee's recommendation concerning National Parks, the recreational value of the South Downs as a potential National Park had been considerably reduced by extensive cultivation and that the designation as a National Park was no longer appropriate. The ploughing continued, albeit at a slower rate, and the battle shifted to maintaining and strengthening the public rights of way on the Downs.

Designated in 1972, the South Downs Way was a milestone in the history of the South Downs. This ridgeway track along the crest of the Downs was the first long-distance bridleway to be created in Britain, an indication of the fervent interest in the South Downs by local and national interest groups alike. When first designated, it stretched over 80 miles, providing a continuous route, unhampered by ploughed fields, fences or locked gates. The route is now 100 miles long, having been extended west to Winchester.

AONB Designation

It took nine years of negotiation between the National Parks Commission, local landowners and 'open air' organisations to achieve designation of the Sussex Downs as an AONB in 1966. The County Councils of East and West Sussex published a Statement of Intent for the Sussex Downs AONB in 1986 which gave a short description of the landscape and identified a range of issues likely to affect that landscape. Following that statement, a Sussex Downs Forum was established with a wide membership that met annually to discuss matters relating to the AONB. The Sussex Downs Conservation Board was established in 1992 to give a voice to the AONB and leadership for practical action on the ground.

In 1991, Countryside Commission policy stated that landscapes designated as AONBs must be of high scenic quality, of national importance and of equal standing to a National Park. The designation therefore recognises that the Sussex Downs is one of England's finest landscapes and that there is national as well as local interest in its conservation and management. It reflects an awareness that such special landscapes are vulnerable to change and, if destroyed, cannot necessarily be recovered.

In addition to the AONB designation, the cliff tops between Seaford and Eastbourne, together with a substantial area of the eastern downland immediately inland, have been defined as a Heritage Coast. Heritage Coast status covers the finest stretches of undeveloped coast in England and Wales. It highlights the need for the effective management and protection of such areas to conserve their natural beauty and to ensure their continued enjoyment by the public.

Most AONBs are administered by local authorities, often through the coordinated action of a joint advisory committee. In 1992 the Sussex Downs Conservation Board came into being as a result of an agreement between the Countryside Commission and the 13 local authorities in the AONB as a joint local government committee. The Board represents a pioneering national experiment in the management of AONBs to see how the protection and management of such areas can be improved within the existing legal framework and with close cooperation between all the agencies concerned. Its progress is being closely monitored by many interested parties, both locally and nationally.

Outstanding Qualities

AONBs are designated for the fine quality of their landscape, in other words, their 'outstanding natural beauty'. They display a range of unusual, unique or exceptional qualities. There is often a combination of factors that give an area its distinctive character and beauty and thus make it outstanding.

The Countryside Commission acknowledges that natural beauty and scenic quality cannot simply be defined as the visual appearance of the countryside alone, but needs to include factors such as landform, vegetation, man-made features, aesthetics and historical and cultural associations. The rarity and representativeness of a landscape as part of the national resource, relative to those in other areas, is also of importance, as is its unique 'sense of place', its accessibility and the public perception of it.

Scenic Qualities

The South Downs is arguably England's best-known, most loved and most used stretch of chalk downland. The classic rolling chalk scenery of the Downs is contrasted with the less well-known landscape of the rugged, wooded greensand ridges of the western Weald. Yet taken as a whole, the strong visual relationships between these two very different landscapes create a sense of overall unity within the AONB. In the north-west the

John Tyler

Beachy Head

John Tyler

Alfriston Church 'the Cathedral of the Downs'

steep summit of Black Down and the wooded greensand ridges act as a counterpoint to the chalk escarpment and both of these contrasting ridges form a strong backdrop to views throughout the area. Together they have an enclosing presence providing a bold setting for a varied range of Sussex landscape types. Further east, the open chalk escarpment seems to stand alone as a dramatic wall, buttressing the chalk landscapes to the south and dominating the wealden plain for miles around.

The AONB has a wealth of beautiful landscape features. Individual highlights include the sweeping panoramas from the chalk escarpment over the patchwork of the Weald; precipitous dazzling chalk cliffs at Beachy Head; the unbroken, curving lines of the open downland; deep, ferny ghylls on the greensand ridges near Chithurst; heathlands veiled in autumn glory - the list could go on and on. This is a landscape with a strong, distinctive character and its strength lies in its variety of different landscape types which together make up the landscape mosaic of the AONB.

The mosaic of different landscape types provides an overall framework for describing and understanding the scenic qualities of the Sussex Downs and the western Weald. Within this broader context, the AONB is enlivened by local characteristic features and individual landmarks such as historic earthworks, prominent clumps of trees, fine buildings and unusual landforms. Burton Millpond, Halnaker Mill, Chanctonbury Ring and Amberley Wild Brooks are all examples of such elements in the landscape which make an important contribution to its individuality and sense of place.

The transitions between the component landscapes of the mosaic may be strong and dramatic, or more of a gentle intermingling of different landscape characteristics. The chalk escarpment is an example of the dramatic. Views from the crest of the downland ridge offer a new dimension to the lowland landscape of south-east England. The greensand ridges are higher still, but the transition between these steep, wooded slopes and the pastures and shaws of the Low Weald is a gentle, muddled, almost imperceptible change from heathy woodland to farmland glades and eventually, gently undulating pastures bordered by winding shaws and hedgerows.

Historical Heritage

Each historical period of human influence has left its mark on the landscape, the traces of earlier cultures rarely being entirely erased by the activities in subsequent years. Thus parts of Stane Street still provide a raised causeway across the western Downs and Cissbury Ring is juxtaposed against the housing estates to the north of Worthing. These well-known, highly visible ancient sites are superb examples of the rich historical heritage of the AONB, but the landscape as a whole has a strong sense of continuity with the past.

The AONB feels ancient. The numerous historical and archæological features - medieval field patterns, Bronze Age barrows, Iron Age hill-forts, strip lynchets - all contribute to this sense of history, but there is more to it than that. The landscape exudes a primitive tranquillity which tends to soothe and strengthen, while at the same time helping us to understand our place within the long track of history.

Wildlife Heritage

The diverse range of landscapes within the AONB harbour some nationally important wildlife habitats: remnants of undisturbed chalk grassland on the Downs; ancient woodland which can be traced back to the original 'wildwood' of Andredsweald; acid and rare chalk heathland; chalk cliff-faces; and the flood meadows and reed-filled ditches of Amberley Wild Brooks. Most are designated Sites of Special Scientific Interest (SSSI) and some of the natural communities represented in the AONB are of international importance.

The rich wildlife heritage of the AONB represents a living record of natural processes and the complex history of human intervention. All the habitats are vulnerable to change and many rely on particular forms of traditional management to maintain their diversity. They are a scarce national resource, increasingly valued as species reservoirs for maintaining biodiversity in a future, more sustainable landscape.

John Tyler

Wild Orchids on Harting Down

Prospects for Change

Despite its protected status, the AONB landscape is under constant pressure for change, some of which is likely to result in its degradation. Agricultural intensification, changes in commercial forestry practice and an overall decline in woodland management, pressures for development, mineral extraction and the growing impact of tourism and recreation may all result in a negative influence if they remain unchecked. The conservation and enhancement of natural beauty, historic heritage and wildlife value of this nationally important landscape depend on action to counter such potentially damaging forces of change. The main prospects for adverse change seem likely to occur from:

- the intense **pressure for development** on the fringes of towns and villages just beyond the boundary of the AONB;
- **new road construction and major road improvements**, both within and on the margins of the AONB;
- **small-scale conversions and building** which, both singly and cumulatively, could lead to the loss of character;
- **changes in the agricultural scene** arising from the decline of traditional methods and changed use of land;
- the continuing **encroachment of scrub** leading to successional woodland which threatens important habitats and the distinctive profile of chalk landforms;
- the **deterioration and even-aged structure of many trees and woodlands**, leading ultimately to a decline in woodland structure and variety, and a threat to their ecological and historical interest; also the lack of management of hedgerows;
- the **potential impact on archæological, cultural and designed landscapes** of a variety of pressures, leading to a degradation of historical features, local variety and sense of place;
- the **increasing numbers of visitors** leading to a damaging change in the balance between the area's unspoilt rural qualities and its traditional tourist role;
- the cumulative effect of **minor road improvements and car-parking provision**, leading to noise, visual intrusion and loss of visual amenity.

A Vision for the Future

The Sussex Downs AONB is a living landscape. In practice, a balance needs to be struck between the conservation of its natural heritage and the needs of those who live and work in it. The economic viability of the community is essential, but if the future AONB landscape is to retain its scenic, historical and ecological value, it must be maintained in a stable condition, particularly in relation to pressures from built development, agricultural change and recreation.

Although these forces for change may be seen as threatening, the landscape of the Sussex Downs which we value so highly today has been fashioned through centuries of change and evolution. Change is not necessarily for the worst and there are many recent examples of change for the better. The ESA Scheme and Farm Woodland Scheme are examples of recent initiatives which have resulted in the expansion of chalk grassland. The establishment of the Sussex Downs Conservation Board (SDCB) represents a substantial investment in this nationally important landscape and offers real opportunities for positive change. SDCB's Management Strategy is designed to ensure the conservation and enhancement of the AONB landscape. This can be achieved in a number of ways:

- **The town and country planning system.** Policies within development plans should promote the conservation of those features that contribute to the character of the AONB and should discourage major development proposals that would detract from it. These policies should be enforced through the development control system.

- **Fostering working relationships.** The future of the AONB's landscape is in many hands: local authorities, farmers and landowners, voluntary organisations, private businesses and individuals. The SDCB provides a formal mechanism to consult freely with all those involved in managing the AONB landscape, taking advice and monitoring results to achieve a valuable flow of information and a mechanism for positive feedback.

- **Working to common aims and objectives.** The preparation of the Sussex Downs AONB Management Strategy is the means to coordinate and guide the process of change effectively. The landscape assessment of the AONB provides a clear understanding of the elements which contribute to the scenic beauty of its landscape and thus a basis for more structured advice and guidance on practical conservation action.

- **Making the best use of resources.** A range of grants and incentives is available to support sympathetic landscape management practices, making them economically viable in the current economic climate. They must be carefully targeted to conserve those areas and landscape features which are most threatened and which contribute most to the special character of the AONB.

- **Improving awareness of the need for conservation of the AONB.** The key is to improve and maintain the flow of information, and guidance for its interpretation, for the broad range of people who live and work in the AONB, as well as the visitors who come to enjoy it on a temporary basis. It is important to establish a common caring attitude towards the landscape as a precious resource, for which everybody has a responsibility, and to encourage as many local people as possible to become involved.

Arundel Castle

© Martin Page

Fulking Escarpment in Winter

A Quintessentially English Landscape

Situated in the most populated corner of the country, the South Downs has a very public face and is arguably one of the most pressurised of our protected landscapes. Yet those who have explored a little will know that this part of Sussex also has many less well-known faces. The AONB harbours a remarkably rich array of beautiful scenery, ranging from the chalk downlands and the ancient bostels, streams and pastures of the spring-line villages at the foot of the scarp, to the swathes of purple heather on the lowland heaths and ancient forests of the greensand ridges. The image of sheep on rolling downland, against a sunny blue sky is for many a symbol of lowland England.

The Sussex Downs Conservation Board is undertaking a pioneering role in effectively managing this important countryside. The Board's role is to ensure that the AONB landscape is protected, conserved and enhanced for future generations.

The appeal of this nationally important landscape is aptly summed up by John Godfrey:

> *England, as Anthony Collet reminds us, is a small country, but full of surprises ... And this is perhaps the essence of the South Downs - the rounded, low hills, the long views over a contented and productive countryside, the sky and the sea, the song of the lark, the winding rivers and their lush valleys, noble castles and brick and flint cottages - full of variety and, yes, of surprises, but set within a homely, comfortable and unthreatening framework, a landscape which it is understandable was chosen in the war years to represent the England that was being fought for.*
>
> *The landscape of the South Downs is essentially picturesque rather than sublime and perhaps for this reason occupies a special place in the affections of Englishmen today, as it has done in the past. Long may it continue to do so.*

References

Allcroft, A H (1924) Downland pathways, Methuen, London.

Arscott, D (1992) The Sussex story, Pomegranate Press, Lewes, Sussex.

Arscott, D (1994) Living Sussex, Pomegranate Press, Lewes, Sussex.

Beckett, A (1909) The spirit of the Downs, Methuen, London.

Belloc, H (1936) The county of Sussex, Cassell, London.

Beningfield, G, Pailthorpe, R, Payne, S (1989) Barclay Wills' the downland shepherds, Alan Sutton, Gloucester, Stroud.

Blackmore, R D (1875) Alice Lorraine: a tale of the South Downs, Sampson Low, London.

Brandon, P (1974) The Sussex landscape, Hodder and Stoughton, London.

Brandon, P (1994) Man in the downland landscape, unpublished manuscript.

Brandon, P and Short, B (1990) The South East from AD 1,000, Longman, Harlow, Essex

Burke, J (1974) Sussex, Batsford, London.

Camden, W (1586 1st edition) Britannia, R. Newbury, London, later editions include Camden's Britannia (1730), E and R Nutt, T Cox, London.

Cobbett, W (1853) Rural rides, reprinted (1957), Everyman, London.

Cochrane, J (1977) Rudyard Kipling: selected verse, new edition (1983), Penguin, London.

Cleere, H and Crossley, D (1985) The iron industry of the Weald, Leicester University Press, Leicestershire.

Coleman, R (1981) Downland: a farm and a village, Viking Penguin, New York.

Countryside Commission (1991) Areas of Outstanding Natural Beauty: a policy statement, CCP 302, Countryside Commission, Cheltenham, Gloucestershire.

Countryside Commission (1993) Landscape assessment guidance, CCP 423, Countryside Commission, Cheltenham, Gloucestershire.

Countryside Commission (1994) Overhead electricity lines: reducing the impact, CCP 454, Countryside Commission, Cheltenham, Gloucestershire.

Curwen, E C (1937) The archæology of Sussex, new edition (1954), Methuen, London.

Darby, B (1975) View of Sussex, new edition (1982), Robert Hale, London.

Defoe, D (1724-1727) A tour through the whole island of Great Britain, new edition (1971), Penguin, London.

East Sussex Woodlands Forum (1990) A trees and woodland strategy for East Sussex, East Sussex Woodland Forum, East Sussex County Council, Lewes, East Sussex.

Edwardes, T (1911) Neighbourhood - a year's life in and about an English village, Methuen, London.

Galsworthy J (1922) The Forsyte Saga, many subsequent editions, Heinemann, London.

Gibbons, S (1932) Cold comfort farm, Longman, Harlow, Essex.

Gilpin, W (1st edition 1804, subsequent edition 1973, out of print) Observations on the coasts of Hampshire, Sussex and Kent, Richmond Publishing Co, Slough, Berks.

Godfrey, J (1995) The history of the South Downs. Downs Country, May/June & July/August, Downs Country Publications, Stedham, Midhurst, West Sussex.

Hampson, F (1987) Literary prizes accepted before afternoon milking. Sussex Life, March, Brownhart Publishing Ltd, Worthing, Sussex.

Hoskins, W G (1977) The making of the English landscape, Hodder and Stoughton, London.

Hudson, W H (1899) Nature in downland, Longman, Harlow, Essex.

Jeffries, R (1883) Nature near London, many subsequent editions, Chatto and Windus, London.

Jeffries, R (1889) Field and hedgerow, many subsequent editions, Longman, Harlow, Essex.

Lowerson, J (1980) A short history of Sussex, Dawson, Folkestone, Kent.

Lucas, E V (1904) The highways and byways of Sussex, Macmillan, London.

Masefield, R (1987) Brimstone, Heinemann, London.

Masefield, R (1982) Chalk Hill Blue, Heinemann, London.

Masefield, R (1988) Painted Lady, Heinemann, London.

Merrifield (1860) Natural history of Brighton, W. Pearce, Brighton, Sussex.

Ministry of Agriculture, Fisheries and Food (1990) Environmentally Sensitive Areas: South Downs, landscape assessment for monitoring, HMSO, London.

Munroe, J (1986) Philip Wilson Steer 1860-1942: paintings and watercolours, Arts Council of Great Britain, London.

Radcliffe, A (1826) A memoir of the author, prefix to Gaston de Blondeville, Henry Colburn, London.

Rose, F (1992) Report on the remaining heathlands of West Sussex, 1991-92, West Sussex Heathland Forum, published by West Sussex County Council, Chichester, West Sussex.

Roy, I (1992) Sussex, Alan Sutton, Stroud, Gloucestershire.

Sparks, B W (1949) The denudation chronology of the dip-slope of the South Downs. Proceedings of the Geologists' Association, Vol. 60, Part 3, pp. 165-215, Geological Society, Bath, Avon.

Sparks, B W (1971) Rocks and relief, Longman, London.

Sparks, B W (1981) (first edn 1972) Geomorphology, Longman, London.

South Bank Centre (1987) The experience of landscape: paintings, drawings and photographs from the Arts Council Collection, South Bank Centre, London.

Sussex Downs Conservation Board (1994) An archæological strategy for the Sussex Downs AONB, unpublished.

Sussex Downs Conservation Board (1995) A landscape assessment of the Sussex Downs AONB, Sussex Downs Conservation Board, Storrington, West Sussex.

Sussex Downs Conservation Board (1994) The statutory planning context: development and minerals, internal working paper.

Sussex Downs Conservation Board and Sussex Wildlife Trust (1994) An overview of the ecological interest of the Sussex Downs AONB, unpublished.

Titman, L (1942) John Galsworthy in Sussex. Sussex County Magazine, Vol. 16, pp. 227-228, Sussex Printers Ltd, Eastbourne, Worthing and London.

Trollope, A (1882) Kept in the dark, Chatto and Windus, London.

Trollope, A (1882) The fixed period, Blackwood and Sons, Edinburgh and London.

Trollope, A (1946) Mr. Scarbrough's family, Oxford University Press, London.

Trollope, A (1936) An old man's love, Oxford University Press, London.

Trollope, A (1883) The land leaguers, Chatto and Windus, London.

Trustees of the British Museum (1985) British landscape watercolours 1600-1860, British Museum Publications Ltd, London.

University of Sussex, The Geography Editorial Committee (1983) Sussex: environment, landscape and society, Alan Sutton, Gloucester, Gloucestershire.

West Sussex County Council (1993) West Sussex literary, musical and artistic links, West Sussex County Council, Chichester, West Sussex.

West Sussex County Council (1994) West Sussex landscape assessment, unpublished.

White, G (1789) The natural history and antiquities of Selbourne, T. Bensley, London.

Woolf, V (1953) Between the acts, Penguin Books, London.

Woolf, V (1963) Diaries of Virginia Woolf, Hogarth Press, London.

Woodford, C (1984) (first edn. 1972) Portrait of Sussex, Robert Hale, London.

Young, Rev. Arthur (1813),General view of the agriculture of the county of Sussex, reprinted (1970), David and Charles, Newton Abbott.

Acknowledgements

We would like to thank the following for their help and advice in the production of this document:
Robert Tregay; Kate Collins; John Godfrey; Valerie Porter; Dr. Peter Brandon; Dr. Rendell Williams; Dr. Brian Short; Martin Hayes and Trevor Seddon.

Aerial photographs
© East Sussex County Council and West Sussex County Council

The Countryside Commission aims to make sure that the English countryside is protected, and can be used and enjoyed now and in the future.

The objectives of the Sussex Downs Conservation Board are:

- to protect, conserve and enhance the natural beauty and amenity of the Sussex Downs Area of Outstanding Natural Beauty (AONB), including its physical, ecological and cultural landscape;
- to promote the quiet informal enjoyment of the Sussex Downs AONB by the general public but only so far as is consistent with the first objective;
- generally to promote sustainable forms of economic and social development, especially working with farmers and landowners to encourage land management which supports the two objectives above.

AONB landscape assessments:

Title	Ref	Year	Price
The New Forest landscape	CCP 220	1987	£5.00
The Blackdown Hills landscape	CCP 258	1989	£5.00
The Cambrian Mountains landscape	CCP 293	1990	£6.50
The Cotswold landscape	CCP 294	1990	£6.50
The North Pennines landscape	CCP 318	1991	£7.00
The Nidderdale landscape	CCP 330	1991	£7.00
The East Hampshire landscape	CCP 358	1991	£7.00
The Tamar Valley landscape	CCP 364	1992	£6.00
The Chichester Harbour landscape	CCP 381	1992	£7.00
The Chilterns landscape	CCP 392	1992	£7.50
The Forest of Bowland landscape	CCP 399	1992	£7.50
The South Devon landscape	CCP 404	1992	£7.00
The Suffolk Coasts & Heaths landscape	CCP 406	1993	£7.50
The Shropshire Hills landscape	CCP 407	1993	£7.50
The Lincolnshire Wolds landscape	CCP 414	1993	£7.00
The Dorset Downs, Heaths & Coast landscape	CCP 424	1993	£7.50
The Malvern Hills landscape	CCP 425	1993	£7.50
The East Devon landscape	CCP 442	1994	£8.00
The Isle of Wight landscape	CCP 448	1994	£8.00
The High Weald landscape	CCP 466	1994	£8.00
The Howardian Hills landscape	CCP 474	1995	£8.00
The Cranborne Chase & West Wiltshire Downs landscape	CCP 465	1995	£8.00
The Cannock Chase landscape	CCP 469	1994	£8.00
The Solway Coast landscape	CCP 478	1995	£8.00
The Kent Downs landscape	CCP 479	1995	£8.00
The Norfolk Coast landscape	CCP 486	1995	£8.00

They can be obtained from the Countryside Commission Postal Sales, PO Box 24, Walgrave, Northampton NN6 9TL. Tel: 01604 781848.